Population and Community Biology

RISK ASSESSMENT IN CONSERVATION BIOLOGY

Population and Community Biology Series

Principal Editor

M. B. Usher
Chief Scientific Adviser and Director of Research and Advisory Services,
Scottish Natural Heritage, UK

Editors

D. L. DeAngelis
Department of Biology, University of Florida, USA

B. F. J. Manly
Director, Centre for Applications of Statistics and Mathematics,
University of Otago, New Zealand

The study of both populations and communities is central to the science of ecology. This series of books explores many facets of population biology and the processes that determine the structure and dynamics of communities. Although individual authors are given freedom to develop their subjects in their own way these books are scientifically rigorous and a quantitative approach to analysing population and community phenomena is often used.

Already published

4. **The Statistics of Natural Selection**
   B.F.J. Manly (1985) 484pp. Hb/Pb.
   Hb out of print.

5. **Multivariate Analysis of Ecological Communities**
   P. Digby & R. Kempton (1987)
   206pp. Hb/Pb. Hb out of print.

6. **Competition**
   P. Keddy (1989) 202pp. Hb/Pb. Hb out of print.

7. **Stage-Structured Populations: Sampling, Analysis and Simulation**
   B.F.J. Manly (1990) 200pp. Hb.

8. **Habitat Structure: The Physical Arrangement of Objects in Space**
   S.S. Bell, E.D. McCoy & H.R. Mushinsky (1991) 452pp. Hb.

9. **Dynamics of Nutrient Cycling and Food Webs**
   D.L. DeAngelis (1992) 285pp. Pb.

10. **Analytical Population Dynamics**
    T. Royama (1992) Hb. (1996) Pb. 387pp.

11. **Plant Succession: Theory and Prediction**
    D.C. Glenn-Lewin, R.K. Peet & T.T. Veblen (1992) 361pp. Hb.

12. **Risk Assessment in Conservation Biology**
    M.A. Burgman, S. Ferson & R. Akcakaya (1993) 324pp. Hb.

13. **Rarity**
    K.Gaston (1994) 192pp. Hb/Pb.

14. **Fire and Plants**
    W.J. Bond & B.W. van Wilgen (1996) ca. 272pp. Hb.

RISK ASSESSMENT IN CONSERVATION BIOLOGY

M. A. Burgman

*School of Forestry, University of
Melbourne, Australia*

S. Ferson

*Applied Biomathematics,
New York, USA*

H. R. Akçakaya

*Applied Biomathematics,
New York, USA*

CHAPMAN & HALL

London · Glasgow · Weinheim · New York · Tokyo · Melbourne · Madras

Published by Chapman & Hall, 2-6 Boundary Row, London SE1 8HN, UK

Chapman & Hall, 2-6 Boundary Row, London SE1 8HN, UK

Blackie Academic & Professional, Wester Cleddens Road, Bishopbriggs, Glasgow G64 2NZ, UK

Chapman & Hall GmbH, Pappelallee 3, 69469 Weinheim, Germany

Chapman & Hall USA., 115 Fifth Avenue, New York, NY 10003, USA

Chapman & Hall Japan, ITP-Japan, Kyowa Building, 3F, 2-2-1 Hirakawacho, Chiyoda-ku, Tokyo 102, Japan

Chapman & Hall Australia, 102 Dodds Street, South Melbourne, Victoria 3205, Australia

Chapman & Hall India, R. Seshadri, 32 Second Main Road, CIT East, Madras 600 035, India

First edition 1993
Reprinted 1994, 1996

© 1993 M.A. Burgman, S. Ferson, H.R. Akçakaya

Typeset in 10/12pt Times by Best-set Typesetter Ltd, Hong Kong
Printed in Great Britain by St Edmundsbury Press, Bury St Edmunds, Suffolk

ISBN 0 412 35030 0

A Catalogue record for this book is available from the British Library

Library of Congress Cataloging-in-Publication Data
Burgman, Mark A.
    Risk assessment in conservation biology/M.A.Burgman, S.Ferson
H.R.Akçakaya
    p.   cm.- (Population and community biology series: 12)
Includes bibliographic references and index.
ISBN 0-412-35030-0 (HB)
1. Population biology - mathematical models. 2. Risk assessment -
    Mathematical models. 3. Population biology - Research -
    Methodology. 4. Wildlife conservation. I. Ferson, S.
    II. Akçakaya, H.R. III. Title. IV. Series
    QH352.B87   1992   92-5461
    574.5'248'011 - dc20   CIP

∞ Printed on permanent acid-free text paper, manufactured in accordance with
ANSI/NISO Z39.48-1992 and ANSI/NISO Z39.48-1984 (Permanence of Paper).

Contents

Preface

In the last decade, ecologists have synthesized methods from demography, genetics, wildlife management and other fields to form the science of conservation biology. One of the central approaches of conservation biology is population viability analysis. Such analyses are not based on a single methodology. Rather, they are a collection of methods applied with the aim of evaluating all those factors that impinge on the persistence of a species. Population viability analysis involves building models of populations to evaluate these factors. To read and profit from the vast majority of books about modelling, with a few notable exceptions, one must be a competent mathematician. Lately, books have appeared oriented towards biologists that explain the fundamentals of model building and the principles of structured populations.

This is a book about how to build models. The models we develop are of a particular type, namely stochastic models for risk assessment. To create realistic models of natural populations, it is necessary to understand stochasticity and the important role it plays in natural systems. To use stochastic models to solve problems in wildlife management, we apply the principles of risk analysis. We want to make these models accessible to the people who need them, conservation biologists and wildlife managers.

We are concerned largely with the risks of population decline. Although many of the methods we use are common to the general field of population dynamics, this concern lends a slant to our questions. This is reflected in the way we build models, in the sorts of questions we consider to be important, the kinds of data we need, how data are interpreted and incorporated into our models.

Wildlife managers provide assessments of the likely impact of a management plan or proposed development on natural populations. Often, such assessments are sought when salient facts such as rates of reproduction and survival and their dependence on age, interactions with other species and so on, are unavailable or incomplete. We work through a series of examples that show how valuable insights can be obtained from models of populations for which data are scarce, making use of whatever information is available without presuming too much and we proceed through to models of populations about which we know a great deal.

This book is not a comprehensive treatment of the subject of models for risk assessment. Such models may become quite complex, especially in circumstances where the biology of a species is very well understood. Here, we treat the main components of model building that one might need for natural populations and concentrate on those aspects of modelling that are particularly pertinent to risk assessment. Thus, in the chapter on structured models, there are numerous aspects of complex structures, parameter estimation and so on to which we make only brief reference. In the chapter on conservation genetics, we have chosen to ignore pedigree analysis, a tool of particular importance to those working on captive and very closely monitored populations; it is a subject that deserves a book in its own right. Our models address only single populations or collections of populations from a single species. We ignore the dynamics of species interactions and such topics as predation and interspecific competition even though they may be important to conservation.

Our main objectives are to stimulate interest in the subject of conservation biology in general and risk assessment in particular, and to continue to bridge the gap between modelling and biology. The most valuable aspect of building models is not the predictions they make, but the process of model building itself. This formal approach to setting out ideas and evaluating their consequences is useful foremost for its rigour, and the process sometimes throws light in the most unexpected places.

M.A.B.
S.F.
H.R.A.

Acknowledgements

We have many people to thank for reading various parts of the book and offering data, critical comments and suggestions: Roger Arditi (University of Lausanne), Rolf Beilharz (University of Melbourne), David Coates (Western Australian Department of Conservation and Land Management), Dick Frankham (Macquarie University), Bill Hampel (University of Melbourne), John Hannagan (University of Melbourne), Steven Hopper (Western Australian Department of Conservation and Land Management), Claire Layman (University of Melbourne), David Lindenmayer (Victorian Department of Conservation and Environment), Hugh Possingham (University of Adelaide), and Nigel Turvey (Shell Indonesia). Michael Usher, the principal editor of the *Population and Community Biology Series* in which this book appears, read the entire manuscript twice and offered numerous constructive comments. Whatever errors remain should be attributed to the authors.

The approach to risk assessment outlined in this book owes much to the pioneering work of Lev Ginzburg (State University of New York at Stony Brook) and his collegues.

Claire Layman prepared many of the figures. The artwork at the start of each chapter representing various animals and plants is hers, with some help from Pat Layman. The paintings are based on photographs and illustrations from the following sources:

Bald ibis (Chapter 1): photograph by Reha Akçakaya.
White rhinoceros (Chapter 2): photograph by Nature Agence Photographique in D. MacDonald (ed.), *The Encyclopaedia of Mammals*: 2, George Allen and Unwin, pp. ii–iii.
Larch budmoth (Chapter 3): illustration by F. Hrozinka and B. Stary in V. Novak (1976) *Atlas of Insects Harmful to Forests*, Elsevier, Amsterdam, p. 41.
Loggerhead turtle (Chapter 4): photograph by Six, in M. Mlynarski and H. Wermuth (1975) The Turtles, *Grzimek's Animal Life Encyclopaedia*, Volume 6, *Reptiles*, Chapter 3, Van Nostrand Reinhold, New York, p. 90.
Gorilla (Chapter 5): photograph by W. Suschitzky in Richard Carrington and the Editors of Time Life, *The Mammals*, Time Life International, London, p. 178.
Matchstick banksia (Chapter 6): photograph by B. and B. Wells (1988) Rare and endangered. *Australian Natural History*, **22**, 354–5.

1 A framework for risk assessment

One of the fundamental tasks of conservation biologists and wildlife managers is to keep as many species as possible from becoming extinct for as long as possible. The abundance of populations change because the number of births in a given period of time rarely equals the number of deaths. The role of people responsible for these populations is to intervene in processes that influence the number of births and deaths in a population, especially when species are close to extinction.

Often, there will be more than one possible course of action at our disposal. For a particular population, certain factors in the environment and in the behaviour of individuals will be important in determining how many survive and produce offspring in any one year. Some of these factors are independent of one another, while others interact.

One approach to making quantitative predictions would be to set up a mathematical model which is a faithful one-to-one reflection of all the processes that we observe. We would follow the fate of each individual through time, and be able to relate the details of growth, reproduction, survival and migration of individuals to components of the environment such as competition, predation and the weather. However, in most circumstances we would not have enough information to be able to characterize all the processes, many of which may not impinge on any interesting question we care to ask, and the model would rapidly become so large and complex as to be unmanageable (Levins, 1966).

A model is a simplification; an abstraction of how we think nature operates. We build models to answer specific questions and, in doing so, incorporate features and processes we see as important and omit those that appear superfluous. To some extent, the purpose of a model will ordain its structure, content and complexity. When the purpose is to make qualitative predictions (such as: will the population increase or decrease?), one should sacrifice precision to realism and generality. Precision becomes important if quantitative answers are needed (Levins, 1966; Starfield and Bleloch, 1986, pp. 1–15).

A mathematical model is a set of equations representing a system. We are interested in mathematical models for biological populations. The advantage in building a formal model is that it forces us to be rigorous in our thinking. It allows us to test the properties of our ideas, and to explore the logical consequences of what we believe to be true.

Models, and by association the people who make them, fall into disrepute when the predictions they make are purported to be true and turn out in practice to be incorrect. The predictions of even the simplest models are often incorrect. Our model is our hypothesis for the dynamics of a species; the predictions we make are tentative and while we expect them to come about, they should always be regarded with scepticism. If we need to work with them to make decisions, we should remember that they are only the best we can do under the current state of our knowledge.

We should be prepared for surprises. It is our responsibility, as people who form models, to make those who rely on them aware of the model's pitfalls and assumptions, and the degree of reliability of its predictions. The predictions may be correct because the model adequately describes the dynamics of a population. On the other hand, any agreement between observations and predictions may be entirely fortuitous. Even if a model makes correct predictions in one set of circumstances we should continue to be sceptical.

Thus, our perspective of models should be, 'if we assume such and such to be true, this is what we would expect'. Then, we look for information that will contradict our predictions. Having found it, we reformulate the model to account for the new knowledge, thereby revising our ideas on how nature works. We can do this by changing the values of parameters in the model, or by changing the way in which the model works, which is to say, by changing the structure of the model.

One of the purposes of a model of a natural population is to help answer questions about the population. There are several common questions in conservation biology. For example:

How will the abundance of a population change in the future?
How long will the population persist?
How long will it take for the population to recover from poaching?
What is the risk the population will become extinct?
Will captive breeding and translocation be worth while?
How should a stocking programme be managed?
Where should reintroductions be planned to optimize results?
Is it better to spend resources to combat poaching or to enhance or extend habitat?

These questions and others like them can be phrased and answered through the medium of mathematical models. The models can be studied to reveal essential details about the natural systems they represent. This book consists mostly of examples of how questions in conservation biology can be formulated and answered using models.

A common way of considering these kinds of questions in the past has been to estimate the rate at which a population grows, or the rate at which it will grow under some management practice. For conservation questions about managed or harvested populations (such as deer or whales) this approach has led to the use of measures such as maximum sustainable yield. Usually, this is taken to be the maximum number of individuals that may be removed from a population while ensuring that the population will persist indefinitely. But no population persists indefinitely. Estimates of population growth rates, population size or maximum sustainable yield are presumed to be measures of the vigour of a population and methodological tools based on these and similar measures have been used widely by wildlife managers. While these approaches deserve discussion, we shall argue in this book that they are inadequate to answer the fundamental questions posed by conservation biology.

In their place, we propose a probabilistic framework in which models are used to assess short-term risks rather than to make precise predictions about the future of a population. In this book, we develop models to account for the inherent uncertainty of biological systems, an approach that is termed risk assessment. Our arguments will be explained through examples in the chapters that follow.

Short-term, probabilistic approaches are relevant and useful for conservation because natural populations live in uncertain environments. We cannot predict exactly the outcome of the host of factors that impinge on any population. The best that can be done is to estimate the chances of particular outcomes based on the variations observed in the past and any mechanistic understanding of the processes that control change in the population.

The variation in a population's dynamics that results from random fluctuations in natural phenomena can be quite dramatic. If we ignore these random fluctuations by using measures of the asymptotic (very long term) behaviour of the population, we miss the fact that the behaviour of the population in the short term may be important, perhaps deciding the fate of the population before the long-term behaviour becomes established. For instance, the abundance a population may eventually reach is irrelevant if, because of some chance event, the last individual dies next year. In general, long-term predictions of the future of a population lose relevance if chances of extinction are high, or if fluctuations in environmental conditions keep the population at relatively low densities.

Look again at the list of questions. Notice that they fall into two broad categories. The first four are questions about the status of a population. The latter four are about management strategies. We distinguish between the two kinds, calling the first *assessment* questions and the second *management* questions. The two are not always distinct but assessments generally call for study of what is likely to happen in the future while management is concerned with the consequences of making a particular decision that affects the species. The techniques one would use for answering assessment questions differ from those for management questions. And, of course, the indices and population measures relevant for conservation issues and the decision criteria used to resolve management questions depend on the nature of the question as well as details of the particular population.

The approach a conservation biologist takes to solve problems is determined by several factors. These include the question to be addressed, the kind and availability of information about a species, the biases and educational background of the researchers, and the agenda of the institution that funds the research, among others. Confronted by a declining population of an endangered species, some will focus on the genetics of inbreeding depression, some on the minimum population size governed by the intrinsic variability of the population dynamics, others on the effect of environmental fluctuations on rates of reproduction and survival, or the effect of habitat fragmentation on migration patterns. Above all, the approach one uses is determined by the scientific intuition of the researchers and their perception of what fundamental biological factors are important for the problem at hand, even within the quantitative framework outlined in this book.

This book is organized into chapters that describe the important approaches that are commonly used in conservation biology and wildlife management. In the rest of this chapter, we discuss the biology, and terminology, of extinction. We also discuss how uncertainty plays an intrinsic and unavoidable role in modelling. We introduce the notion of **risk** as a way to make predictions in the face of uncertainty. In Chapter 2 we discuss elementary (linear and unstructured) models of population growth. We use the white rhinoceros as an example to show how one can begin to build a model with only limited data. Chapter 3 expands these models and considers the various kinds of non-linearity induced by **density dependence**. Chapters 4 and 5 cover demographically and spatially **structured** models. Chapter 6 addresses **genetic** factors relevant to population dynamic models. Chapter 7 outlines how the results of these analyses may be implemented to solve conservation problems.

We cannot give a prescription to tell whether an age-structured, a stage-structured or a spatially structured model should be used, or whether the model should focus on demography, habitat or genetics. But in each chapter, we shall point out the particular cases that may require application of one of these approaches, and then describe the tools or models one may develop to answer different kinds of questions.

All of our models address the dynamics of single species. The models usually ignore community dynamics and interspecific interactions such as predation, competition and mutualism. This is not to suggest that species interactions are not important, but this topic deserves a book in its own right. We return to the subject briefly in Chapter 7. Our main theses about short-term probabilistic approaches are relevant to models that include multiple species and their interactions.

The approach we advocate for the measurement of impacts for wildlife management and conservation is to estimate the risks of unwanted events. Population models that include random parameters are called **stochastic** models. They allow us to evaluate risks in terms of population abundance, accounting for the inherent unpredictability of biological systems. If stochastic models are to be applied widely to make useful predictions, it is important that biologists build them. Models are useful only in so far as they help to order our ideas, and suggest relationships that may not be immediately obvious. They will point out areas where there is a lack of data. They may, if sufficiently robust, and properly validated and verified, provide trustworthy predictions of future events.

The mathematics involved in finding solutions for these models is daunting, even to mathematicians. This is especially true when a model has non-linear elements and the majority of intuitively appealing models have them. Most wildlife managers and conservation biologists have, at best, a passing acquaintance with stochastic, non-linear mathematics. The problem, then, is to make the methodology of models, and of stochastic models in particular, accessible to biologists. In this book, we attempt to solve the problem by constructing models that can be solved by Monte Carlo computer simulation.

As dynamic models usually can be expressed either in discrete or continuous time, most of the topics in population biology have dual treatments in the literature. As most (although not all) of the models we discuss will be formulated in discrete time, we shall not be worrying about derivatives and limits and so on. The mathematics is often easier in the

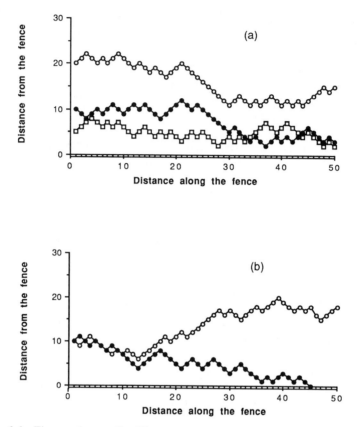

Figure 1.1   Five random walks. The sequences were generated by tossing a coin. The walks in (a) were begun five (squares), ten (solid circles), and twenty (open circles) steps from the fence. Both walks in (b) were begun ten steps from the fence.

discrete case and discrete-time models are generally easier to implement in computer programs.

In several examples we shall describe the actual algorithms used to apply the model on a personal computer. This is unlike the approaches used in many other books that give clean mathematical equations and leave the reader to translate them into computer programs if the need arises. We assume some familiarity with statistics, linear algebra and computer programming, or at least a lack of fear of these subjects. The algorithms we give are not the only way to implement any given model, but they are concrete examples that can help to understand how a model behaves. They can also clear up little ambiguities that can creep into unaccompanied equations (such as, does an index start at zero or one?). We hope that the algorithms will help to make the case that these models are not just sterile academic amusements but that they can be useful in routine work in conservation biology.

1.1   A PROBABILISTIC FRAMEWORK

In many ways, change of population size with time is like a random walk, a mathematical concept that we can treat more literally. Consider yourself walking parallel to a boundary fence. At every third step you take a pace sideways, either towards or away from the fence. You decide which direction to take after flipping a coin. If it is heads you move towards the fence and if it is tails you move away (Figure 1.1).

The chance of crossing the fence within, say, the first 50 steps is greater if you begin close to the fence than if you begin very far away. The precipitous fall in the middle of the upper line in Figure 1.1(*a*) would have led the walk begun at five steps from the fence to travel across the fence. Even though the circumstances surrounding each walk are the same, some might wander away from the fence, avoiding it indefinitely, while others will eventually hit it (Figure 1.1(*b*)). The same starting point can lead to very different outcomes.

The environment is full of unpredictable events like bad weather, disease or the discovery of untapped food supplies. Such events act much like a coin toss in determining the increase or decrease of a population from one season to the next. Thus, it seems likely that there will exist a relationship between the number of individuals in a population or a species and the probability of crossing some population threshold, including the extinction of the species.

We can consider the fence to be an absorbing boundary. That is, it is all right to wander randomly, as long as you avoid the fence. If you don't, you cannot move away from it again. For a natural population, extinction is an absorbing boundary. If ever the population size falls to zero it cannot recover, no matter how favourable are the environmental conditions that follow.

Variability is ubiquitous in biological systems. All populations are engaged in a random walk. In some, variability dominates the dynamics of the population, in others deterministic pressures are far more important and randomness plays a secondary role. The most appropriate type of model depends on the relative importance of these features.

The complexity of the model we build rests on a different set of constraints. Models are most often limited by the amount and quality of available data. This is true not only of population dynamics, but of many other fields for which mathematical models have practical applications. For example, the physical processes determining the temperature at different points in a body of standing water are well understood in some cases and very precise and reliable predictions may be made when relevant data are available (Imberger, 1985). The resolution of a model, its detail and scope, must be a result of a balance between its purpose, the available data and our level of understanding of the system.

Data are required for the estimation of parameters and for validation so that we may begin to use the model with some degree of confidence.

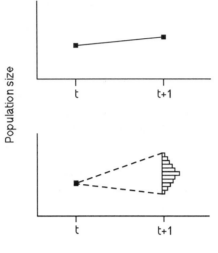

Time

Figure 1.2 The difference between deterministic and stochastic models. The upper diagram results from a deterministic model and represents population growth from time t to time $t + 1$. The population size is predicted exactly. The lower diagram results from a stochastic model. It predicts that the population size at time $t + 1$ could occur anywhere within some set of values. The different possible outcomes have different probabilities of occurring.

The challenge of models for natural populations is that, for the vast majority of problems, data are scarce (Holling, 1978). For that reason, if for no other, the majority of models for natural populations will have relatively few parameters and few complicated processes. Understanding is the conceptual clarity with which we see the processes underlying the model. With understanding, we can see the relationships that characterize a process and be confident that a model's predictions are the result of meaningful biological dynamics.

1.1.1   A simple stochastic model

Stochasticity is statistical uncertainty. To include aspects of stochasticity in a model, it is necessary first to construct a deterministic model to characterize the processes important for the dynamics of a population. Models of population growth usually describe the way in which a population develops from past states. The state of the population at any time is given by a set of variables representing such things as age structure, population size, reproductive rates, environmental conditions, and so on. A discrete-time deterministic model for the size of a population may take the form

$$N_{t+1} = RN_t \qquad\qquad (1.1)$$

which says that the population size, N, at time $t + 1$ is equal to the population size N at time t, multiplied by a growth rate, R. The values for the various parameters are specified exactly. For example, we may say that the population increases by a factor of 0.1 per year, so that R equals 1.1 and one unit of time is a year.

The model represented by Equation (1.1) is fine as long as the growth rate, R, remains constant from one time to the next. However, growth of a population depends on many uncertain factors such as weather and food availability. We could represent this in a model by allowing R to vary in response to random variations in the environment, that is, by allowing the population size at time $t + 1$ to depend on parameters that change randomly. The result of this approach is a probability distribution wherein there is a set of possible sizes for the population at time $t + 1$. The model specifies the probability that N_{t+1} belongs to the set of possible values for the size of the population, given N_t (Figure 1.2; see Chesson, 1978). In such a model, we may estimate the average amount by which a population increases each year (say, 0.1) and also estimate the confidence limits for the average amount of increase.

Although the population itself will finish at a single abundance by time $t + 1$, we do not know which it will be. Furthermore, each possible size at time $t + 1$ will explode into its own distribution at time $t + 2$.

Assume a population of orchids is made up of 100 individuals. Over the past ten years, the population has increased on average by a factor of 0.1 per year. We want to know the population size in six years' time. One way to predict it is to use Equation (1.1) with R equal to 1.1, giving

$$
\begin{aligned}
N_6 &= 100 * 1.1 * 1.1 * 1.1 * 1.1 * 1.1 * 1.1 \\
&= 100 * (1.1)^6 \\
&= 100 * 1.77 \\
&= 177
\end{aligned}
$$

We know that the population depends on good winter rains for germination and that in some years the population does better than avarage and in other years it does worse. We could typify this aspect of the population's dynamics by saying a good year results in an average population increase of 0.2 ($R = 1.2$), while in a poor year the population remains the same size (population increase equals 0 and $R = 1$).

Good and poor years are equally likely with the result that we might expect the avarage rate of increase of the population, $\bar{R}$, to be $(1 + 1.2)/2 = 1.1$. Thus, we would still expect the population size to be close to 177 plants in six years' time. This, in fact, is incorrect. Taking good and bad years in turn, we have

$$
\begin{aligned}
N_6 &= 100 * 1.2 * 1 * 1.2 * 1 * 1.2 * 1 \\
&= 173
\end{aligned}
$$

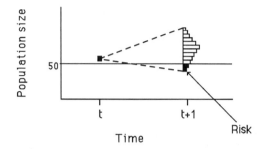

Figure 1.3 A part of the probability distribution falls below the threshold representing an unacceptably small population, predetermined as 50 individuals. The blackened area beneath the line for 50 individuals at time $t + 1$ represents the risk faced by the population. This area, as a proportion of the total area of the distribution at $t + 1$, is the chance that the population size will fall below 50 individuals.

which is not 177. It turns out that the mean rate of growth of a population in which R varies is given by the geometric mean of the growth rates, so that

$$\bar{R} = \sqrt[6]{1 * 1.2 * 1 * 1.2 * 1 * 1.2}$$
$$= 1.095$$

Even if we know the population faces either good or bad years with equal probability, it is not possible at the outset to say whether next year will be good or bad. This uncertainty means we cannot be sure what will happen in the future. At the end of one year the population size could equal 100 or 120, and there is a 50% chance it could be either. At the end of the second year, the population size could equal 100, 120 or 144, each with a different probability of occurring (0.25, 0.5 and 0.25 respectively). Because this example uses an environment with only two states, good and bad, it would be possible to calculate the chances of any outcome using the binomial distribution (see Sokal and Rohlf, 1981).

While the deterministic and stochastic properties of a model are intimately related (Nisbet and Gurney, 1982), the mean result of a stochastic model is not necessarily the same as the prediction of its deterministic analogue (Chesson, 1978). Especially in those cases where a model has strongly non-linear components, predictions of a deterministic and a stochastic model may be qualitatively different. We shall return to this topic in Chapter 3. Stochastic models are not a salve for deterministic systems from which some crucial knowledge is missing. Rather, they may account for a fundamental component of biological systems ignored in deterministic models.

The important point is that the stochastic model provides a probabilistic framework in which to evaluate predictions. It may be that an assumption of randomness implies neglect, or ignorance, of details of the system being modelled, with the result that one is only able to predict the probability that a population will be of a particular size (Nisbet and Gurney, 1982). However, there exist elements of true stochasticity in biological systems that cannot be removed by the collection of additional data (Ginzburg, 1984), such as the inherent unpredictability of fluctuations in the weather and the recombination of genes during meiosis.

1.1.2   Risk assessment

Risk assessment is introduced into stochastic models very easily, at least in concept. Consider the stochastic model demonstrated in Figure 1.2 above. We noted that there is a probability density associated with the prediction for population size at time $t + 1$. Assume that we know a population smaller than, say, 50 individuals is susceptible to inbreeding depression and is not viable. Thus, a population size of 50 is unacceptably small. We want to know what the chances are that our population will be below 50 individuals at time $t + 1$. We add this threshold of 50 individuals to the analysis already completed, giving us Figure 1.3.

We note that part of the distribution falls below the line for 50 individuals. The proportion of the total probability distribution falling below the threshold is equal to the risk faced by the population. Thus, even though the deterministic model tells us the population will increase, and the stochastic model tells us the population will probably increase, there is a measurable risk that there will be an unacceptably small population at $t + 1$. The task of wildlife managers, then, is to implement plans to minimize that risk.

The probabilities generated in the example of the orchid population above allow us to pose different kinds of questions. We might want to know the worst possible outcome for the population: if things go as badly as possible, what will the population size be? Within the limits of our model, the worst that can happen is to have six years in a row of poor rainfall, in which case the population size will be

$$N_6 = 100 * (1.0)^6$$
$$= 100$$

Because the chances of good and poor years each occur with a probability of 0.5, the chance (p) of this coming about is the same as the chance of tossing a coin and getting six heads in a row,

$$p = (0.5)^6$$
$$= 0.015625$$

Thus, we can say there is about a 1.5% chance that the population size will be as low as 100 plants at the end of six years. This statement represents an estimate of the risk the population faces where the event of concern is the population size at the end of six years.

Risk analysis allows us to evaluate the variable components of a population, independently of the mean or the deterministic expectations. For example, we may want to know what difference a 20% decline in population size will make if a population fluctuates naturally around the mean by an order of 50%. Some impacts may have no effect on mean population size but greatly affect variability. Clearly, these impacts will increase the chances of crossing a threshold.

Judgement of risk is commonplace in human society. We judge the acceptability of risks in cars and aeroplanes by what most people are prepared to put up with. The majority consider it a worthwhile risk to fly in an aeroplane, even though there is a chance of dying as a result.

The trouble is that humans are not very good at guessing risks. Zeckhauser and Viscusi (1990) reviewed the literature on human response to risk and concluded that people are often irrational and emotive. We tend to overestimate the likelihood of low probability events (such as death by tornado) and underestimate higher risk levels (such as death by heart disease). Our perception of risk is coloured by such things as how visible the risk is. There is a tendency for people to underestimate risks that they believe they cannot do much to control. Such perceptional biases account for many emotional public responses and eventually will affect decisions made by regulatory organizations and their agents, including politicians and wildlife managers.

Managing risk for natural populations is concerned with allocating resources to wildlife conservation. These resources are scarce because commitment of resources to conservation results in economic trade-offs to the rest of society. If our decisions are emotive, we face the possibility of inefficiency in using resources with the consequent loss of species that might otherwise have been avoided. The best way to address this problem is to apply methods that result in quantitative evaluation of risks for natural populations. We can use these results to underpin management decisions. Thus we might, by consensus, define an acceptable level of risk for the extinction of species. We could then use this benchmark to help allocate conservation resources: an assessment of the risks faced by species will tell us if they are above or below the acceptable level of risk and, if the risks are too high, we shall manage the population in such a way as to reduce the risks.

It is important when dealing with natural populations to recognize that there is never zero risk. Although we may define and consider limits which we set arbitrarily close to zero, we can never, whatever we do, guarantee the survival of a species for any period of time. This point is particularly important because people have a bias towards situations in which they believe there is no risk (Zeckhauser and Viscusi, 1990).

Risk analysis for natural populations involves the calculation of probabilities of adverse events, and it is hard to falsify a probabilistic prediction because the prediction is not exact. When the unexpected happens, to use the words of Ginzburg *et al.* (1990), we can shrug with the immunity of a weather forecaster, claiming we never said the population would not go extinct; we merely said it had a low probability of going extinct. Judgement as to what is, and what is not, an acceptable risk involves a judgement which is by definition anthropocentric. It must inevitably depend to some extent on the economic and cultural value to humans of the species under consideration. It is essential, therefore, to distinguish the tool, risk analysis, from the objective, the conservation and protection of species (Burgman and Neet, 1989). The tool applies equally well to plagues and population explosions of unwanted species such as locusts and rats. Risk analysis permits us to make decisions that modify our actions to reduce the chances that a population will go extinct. The questions remain as to which species to study, and what level of risk of extinction is acceptable. Conservation of species requires the allocation of limited resources, and risk analysis provides one tool for the efficient allocation of those resources.

1.1.3   The language of risk and conservation statistics

Biologists are used to thinking in terms of the central tendency of their data: what is the mean and variance, what are the confidence limits? While these are important concepts and we shall use them, it is just as important to know the distribution of outliers: what is the worst case we might expect, and how likely is it? These latter questions are particularly relevant to people interested in keeping population sizes within pre-determined limits. To look at data or model predictions in this way requires a new vocabulary. We have used some of these terms already and to some extent they foreshadow the methods developed in Chapters 2 and 3 below.

Shaffer (1981) distinguishes between 'systematic pressure' and 'stochastic perturbation' as causes of population extinction. Populations that would persist indefinitely in a constant environment nevertheless face some risk of extinction through variation in vital rates. These populations are the victims of stochasticity. If the long-term growth rate of a population is negative, it will become extinct, no matter how variable the environment. These populations are the victim of systematic pressure, and population decline results from deterministic causes. In fact, Ginzburg *et al.* (1984) observe that if a population has negative mean growth rate, variability may temporarily improve the chances of population persistence.

Risk is the potential, or probability, of an adverse event. **Risk assessment** is defined by Rowe (1977) as the process of obtaining quantitative or qualitative measures of risk levels. When applied to natural populations of plants and animals, risk assessment usually is concerned with the calculation of the likelihood that populations will fall below some specified size. The converse is also important, the risk of population explosions.

All populations live in unpredictable, variable environments. All population sizes vary. Over a given time period, there is a chance that any population will become extinct, or fall below a specified small population size. This is so, even if the population would persist indefinitely in a uniform environment. This chance we term the **background risk** of extinction.

Risk analysis may be used to describe impacts on populations in probabilistic language. By measuring impacts in terms of risks, it is possible to compare impacts against background risks that a population faces in the absence of the impact. **Added risk** is the increase in risk that results from some impact on a natural population, and it provides us with a means of quantifying the impacts of different management practices. Anthropogenic responsibility is limited to added risk (Ferson *et al.*, 1989).

There are several statistics that may be employed to describe the risks faced by natural populations. We can use the expected time to extinction (or the expected persistence time) of a population. Alternatively, we can use the **probability of extinction** of a population. In using the probability of extinction to measure the consequences of our actions, we must consider a time frame, a **time horizon**, over which the potential impact must be measured. The results can be expressed in such terms as 'there is about a 95% probability that the population will survive for at least 50 years', where both the median and the mean time to extinction are larger than 50 years. Probability of extinction as a function of time is often a more informative representation of risk. This is because in using the probability of extinction to measure the consequences of our actions, rather than time to extinction, we include a time frame. In doing so, we account for both the mean and the variability of population growth rates. We recommend this approach in the formulation of models.

Often, assessing the risks faced by a species involves estimating the chances of a population's falling below some specified size. This population size may be thought of as an unacceptably small, critical threshold. It could be the population size at which demographic accidents, behavioural constraints, or genetic factors such as inbreeding depression, begin to play an important role in the continuing survival of a species.

Ginzburg *et al.* (1982) suggested the use of a more flexible term in demographic studies. **Quasiextinction** risk is the chance of crossing some small population threshold, a lower bound that may be unacceptable for conservation, management, economic or aesthetic purposes. The question is usually phrased: 'What are the chances of the population being smaller than the critical population threshold size, N_C, at least once in the next t years, given an initial population size of N_0?' Quasiextinction does not have to refer to a decline to some critical level. As it is defined, it refers to all kinds of declines, the most extreme of which may be

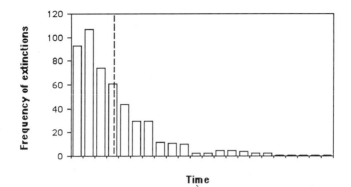

Figure 1.4 Histogram, based on 500 simulation trials, of time to extinction for a population with a ceiling of 100 individuals, an average instantaneous population growth rate of 1.0, and standard deviation in growth of 0.6 (based on the logistic model for population growth). Only environmental factors contributed to the variability in the population growth rate. See the text of Chapter 3 for an explanation of these terms. The average time to extinction, 15.2 years, is shown by the vertical broken line.

demographically critical. Even moderate declines may be important for economic or aesthetic reasons.

In many models developed for conservation biology (e.g. Leigh, 1981; Goodman, 1987a,b; Quinn and Hastings, 1987; Pimm *et al.*, 1988), the variable used to express risk is **time to extinction**. Since the longevity of a species cannot be predicted exactly, the shape of the distribution of the predicted time to extinction is very important. Expected persistence times are not distributed in a regular fashion where the mean, median and mode have the same value. Shaffer (1981; Shaffer and Samson, 1985), for example, points out that while a Yellowstone grizzly bear population has an expected persistence time of 114 years, 56% of these populations will be extinct in periods less than 114 years. It is important to know the mean and the variance of estimates of times to extinction, as well as the type of statistical distribution from which they come.

Goodman (1987a) observes, like Shaffer (1981), that expected times to extinction are strongly skewed to the right. In fact, take almost any model for population growth and the predicted distribution of extinction times will be skewed to the right. In Figure 1.4, the average time to extinction is 15.2 years. However, the times are strongly skewed and the median and the mode values are less than the average. The most frequent time to extinction is just four years, and half the populations are extinct after 11 years. A total of almost 65% of the populations are extinct before the average time.

The times in Figure 1.4 are distributed somewhat like a probability density function termed the exponential distribution, and this type of distribution of expected extinction times is predicted by all models so far explored (for example, see the theoretical studies by Lewontin and Cohen, 1969; Levinton and Ginzburg, 1984; Goodman, 1984, 1987a,b; Lande and Orzack, 1988; and the review by Tuljapurkar, 1989). That is, in most cases we should expect the mode to be less than the mean. Most populations will become extinct at times less than the average. A few populations will persist for much longer than the average. The degree to which the mean time to extinction differs from (i.e. overestimates) the mode depends on the variance. The more variable is the estimate, the greater the difference between mean and mode.

The distribution of extinction times can lead to some counter-intuitive results. There is a model for population growth that we shall explore in Chapters 2 and 3 in which the population grows without limit. If enough time elapses, the population will become very large. Surprisingly, even though we may expect the population to increase, each population in a set of populations growing under the same conditions may be virtually certain to become extinct (Lewontin and Cohen, 1969). Levinton and Ginzburg (1984) discovered that with a similar kind of model, the distribution of extinction times could be so skewed that it does not have a mean.

1.1.4   Types of uncertainty

The events that lead to the extinction of a species may appear in retrospect to be deterministic. For instance, in the example of the bald ibis we detail below, two of the last three wild birds in Turkey were killed by a storm in 1989. There is little chance the storm could have been predicted more than a few days in advance. It would have been much harder to predict how many of the birds would survive the storm. Thus, when faced with predicting the future of an extant population, there will be unknown, and perhaps unknowable, factors that make our predictions uncertain.

The variability that contributes to the random walk in which all populations are engaged may come from different sources. It is possible to classify the different types of variability that impinge on a population. Different classifications have been used by Chesson (1978) and Shaffer (1981, 1987). We use four categories that can unambiguously subsume all types of variation that affect population growth:

1. Phenotypic variation
2. Demographic variation
3. Environmental variation
4. Spatial variation.

Phenotypic variation is the variation between individuals within the same population. These individuals experience the same environmental and demographic conditions and any differences are due to genetic and developmental differences among individuals that result in differences in phenotype. This category includes Shaffer's genetic uncertainty and Chesson's phenotypic variation. Shaffer defines genetic uncertainty as random changes in genetic make-up due to genetic drift, inbreeding or the founder effect which alter the survival and reproductive probabilities of individuals.

Demographic variation is the variation in the average chances of survivorship and reproduction that occur partly because a population is made up of a finite, integer number of individuals. Remember the example of the orchid population above. The population began at 100 plants, and with a growth rate of 1.1 the population will be 110 plants next year. Over the next three years, the population sizes will be 121, 133.1 and 146.41 plants. We have run into a problem. There is no such thing as 0.1 or 0.41 of an orchid. The growth rate we specified is an average based on observations of past populations. What this result says is that in some four-year periods, the population will increase to 146 plants, in others it will increase to 147. On average, over a four-year period, there will be between 46 and 47 more births than deaths in the orchid population. Exactly how many, we cannot be sure.

Many other elements of population structure contribute to this kind of uncertainty. For example, in species with separate sexes, uneven sex ratios may arise by chance and have a detrimental effect on further population increase. This element of uncertainty is called demographic variation, and it arises because real populations are discrete, structured and often quite small. The category was recognized by May (1973a), Chesson (1978) and Shaffer (1981) as important for the dynamics of populations.

Environmental variation is unpredictable change in the environment through time. An obvious example is rainfall. Even in circumstances where we know precisely the average annual rainfall of a location based on records going back centuries, it is notoriously difficult to say if next year will be relatively wet or dry, and even if next week will be rainy or not. It is possible to calculate an average for almost any environmental variable. It is impossible to say what the value of that variable will be at the same time next year. Environmental variation results in fluctuations in population levels because environmental variables affect the number of survivors and the number of offspring in a population. If water is scarce, more juveniles and adults die than if water is plentiful. Shaffer and Chesson both recognize this category as important in determining the dynamics of many populations.

Spatial variation is the variation in environmental conditions between spatially separate patches of habitat, the different conditions experienced by each of several populations. This category corresponds to Chesson's 'between patch variation'. Most species consist of an assemblage of populations that occur in more or less discrete patches of habitat. Only if these patches are identical in composition, close enough that they experience the same environmental conditions, and if there is sufficient migration between them, can we safely ignore patch dynamics. However, in most real populations, at least one of these conditions will be violated. All of the populations will experience some environmental changes in common (such as winter) and some will experience local environmental changes uniquely in a given patch (such as the local waterhole drying out). The pattern of local extinction and recolonization of populations can have profound effects on the risks of extinction of a population.

Shaffer defines another category that we include within environmental variation. He calls it catastrophes and it includes natural events such as floods, fires and droughts. They are different because they have a relatively large effect on the chances of persistence of a population, much greater than those environmental changes that cause the normal year to year fluctuations in population size.

Hilborn (1987) likewise classifies rare and unexpected events and terms them, appropriately, 'surprise'. That does not mean that surprise itself is rare, only that each event is essentially unexpected. Nevertheless, once we have seen Meteorite Crater in Arizona, we should accept the fact that a large meteorite fell there, even though the *a priori* probability of the event is diminishingly small (van Valen, 1984). The terms of reference of Hilborn's definition are a little broader than Shaffer's catastrophes. Surprises include anything we do not expect, anything that is unaccounted for by our model or by our intuition.

1.1.5   The language of population size and distribution

Small populations are more likely to become extinct than large ones. This idea is perceived by conservation biologists when they write lists and develop management programmes for rare species, usually defined as those with small population sizes or restricted geographic ranges.

A common definition of a **population** is a collection of individuals that are sufficiently close geographically that they can find each other and reproduce. In practice, it is any collection of individuals of the same species distributed more or less contiguously. The implicit assumption here is that if individuals are close enough, genes will flow among individuals. Often, biologists in the field are asked to determine the geographic boundaries of a population. The limits of a population depend on the size and lifeform of a species, its mode of reproduction, mode of seed or juvenile dispersal, its habitat specificity, and pattern of distribution within its geographic range.

Subpopulations are parts of a population between which gene flow is limited to some degree, but within which it is reasonable to assume individuals are panmictic (i.e., completely mixed). All of the factors that make it difficult to define the limits of a population are magnified when trying to determine the limits of subpopulations. A **metapopulation** is a

Geographic range							
Widespread				Restricted			
Local Population Abundance							
Somewhere common		Everywhere rare		Somewhere common		Everywhere rare	
Habitat Specificity							
Wide	Narrow	Wide	Narrow	Wide	Narrow	Wide	Narrow
1	2	3	4	5	6	7	8

Figure 1.5 Eight categories of population size and distribution (modified from Rabinowitz *et al.*, 1986). The numbers 1–8 are arbitrary labels for the categories.

set of populations of the same species, usually more or less isolated from one another in discrete patches of spatially separate habitat, that may exchange individuals through migration.

Rarity is an intuitive concept frequently used in conservation and wildlife management, particularly for the evaluation of land for reserves (Usher, 1986). Various aspects have been discussed by Harper (1981), Margules and Usher (1981) and Cody (1986). Several terms related to the concept of rarity have been used interchangeably; the following terms are defined to refer to mutually exclusive characteristics of population size and distribution. These terms were defined and explained by Brown (1984) and Rabinowitz *et al.* (1986). Another introduction to the terminology of conservation biology is provided by Fiedler and Jain (1990).

The concept of rarity can be broken down into three components that represent different aspects of the distribution and abundance of a species, namely **abundance**, **range** and **habitat specificity**. The term 'abundance' refers to the density of individuals within a local area, and the terms **rare** and **common** (or abundant) describe extremes of density. The term 'range' refers to the spatial distribution of a species, and the terms **restricted** (or local) and **widespread** describe extremes of spatial distribution. 'Specificity' refers to the habitat range of a species, and the terms **generalized** (or wide) and **specialized** (or narrow) describe extremes of habitat specificity. Specificity is a measure of the range of ecological conditions, both physical and biotic, under which a species can survive. The limits include all those things (such as temperature, wetness, soil conditions, nutrient status, availability of prey) that preclude the growth and reproduction of a species.

We can add a further term useful in the study of species distributions and abundances. **Ubiquity** describes the frequency of species in samples (i.e., the proportion of samples in which a species is found) and the terms

scarce and **ubiquitous** describe the extremes (Burgman, 1989). These categories do not describe an intrinsic property. Rather, they describe a sampling phenomenon that is a function of a population's abundance, range and specificity. Ubiquity includes an implicit relationship between the size of a sampling unit and the size of the individuals being sampled. Large trees may be scarce in small quadrats but ubiquitous in large quadrats.

Each category in Rabinowitz's classification (Figure 1.5) describes classes in a continuum. The application of these terms depends on the geographical scale of the population relative to the area of interest. Nevertheless, the terms outlined above summarize the related components of population size, distribution, tolerance and frequency recognized in other nomenclatures. For example, Drury (1974) specifies three types of geographic distributions for rare species: (1) species inhabiting stressed sites (in which there are a few individuals wherever there is habitat); (2) widespread but locally infrequent species; and (3) species in large numbers in a few locations. The first category includes rare, specialized species. The second includes widespread, rare species. The third includes common species that may be either scarce or restricted.

1.1.6 Conservation status

When biologists communicate about the risks faced by different species, they use a set of qualitative categories for different degrees of threat. These categories go under the heading of the conservation status of a species. Terminology for conservation status has been developed principally by agencies for conservation and environmental impact assessment. The terms may be found in such publications as he IUCN Red Data Books (see Collar and Stuart, 1985; Munton, 1987) and many other more local publications (see, for example, Leigh *et al.*, 1981).

The terms that describe distribution and abundance are often used interchangeably with those that describe conservation status whereas they describe logically different and, to some degree, independent things. When something is rare it is not necessarily threatened with imminent extinction, just as species that are likely to become extinct in the near future are not ncessarily rare, restricted or specialized. Conservation status embodies estimates of the relative degree of threat faced by difference populations or species.

Superficially, the word 'extinction' has an obvious meaning. However, confusion and bias may arise when the term is applied. It is useful to distinguish between **extinct** species and **presumed extinct** species. Extinct species are those for which there is no doubt that the last member of the species has died. Presumed extinct species are those that have not been definitely located in the wild for the last 50 years. There may be doubt that the species is extinct if, for example, its habitat is remote or inaccessible, its growth form or life history strategies make it difficult to survey, or if there has been a lack of thorough field surveys to establish its absence.

The reason for this distinction is that it is difficult to know the point at which a species should be considered extinct. There are many documented cases of species that were presumed extinct, only to have been found after some considerable time. On the other hand, Diamond (1987) has pointed out that extinction rates will be greatly underestimated, where the flora and fauna are poorly known, if species are considered extinct only after exhaustive field work has failed to locate any member. For these reasons, we differentiate above between the categories 'extinct' and 'presumed extinct'.

Rediscovery of an 'extinct' species
The dibbler (*Parantechinus apicalis*), a small carnivorous marsupial native to Western Australia, was not seen between 1884 and 1967. In 1967 three individuals were trapped in a patch of *Banksia* heathland at Hassell Beach near Albany. Between 1967 and 1985, five populations were found, two on small islands near the coast. While the species is both rare and restricted, and the three populations on the mainland are sparse, the dibbler is very much extant. Fossil deposits indicate the species was once widespread in Western Australia and there is little doubt that land clearing for agriculture, frequent fire, and predation by cats and foxes have contributed to its decline (Dickman, 1986).

It is also worth while to distinguish between **local** and **global** extinction. Local extinction, sometimes called extirpation, refers to the extinction of a single population in a spatially separate patch of its habitat. Global extinction refers to the loss of all members of a species in all of its constituent populations. Since there is often a great deal of local genetic differentiation among populations in a species, loss of a patch can be important.

The conservation status of a species describes the degree of threat a species faces and is related to its chances of extinction. Implicitly, it employs a time scale. A species that has little chance of extinction within ten years may face certain extinction within a century. Two terms are frequently used in conservation literature. **Endangered** species face a high risk of extinction within one or two decades if present causal factors continue to operate. **Vulnerable** species are not currently endangered but they are at risk over longer periods (usually 50 to 100 years) if factors tending to push the species into decline continue to operate. These categories are essentially comparative. Rarely are the risk levels or the time horizons specified (i.e. what is meant by 'a high risk of extinction'

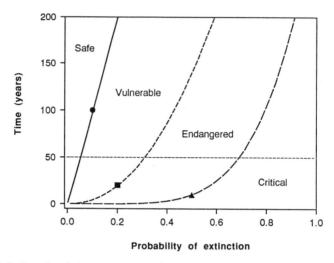

Figure 1.6 Levels of threat as a function of time and probability of extinction. (From Akçakaya, 1992; reproduced with permission of Elsevier, Amsterdam.)

and over what period is this risk evaluated?), largely because species are assigned to different categories without any quantitative analyses. Assignment is based on the experience and intuition of biologists. Different species have different expected lifespans and it may be equally useful to couch a time horizon for conservation status in terms of generations rather than years.

There are a number of other categories to which species are sometimes assigned, depending on the amount of available information or the perceived risks faced by the populations. **Rare, restricted** and **specialized** species are not presently vulnerable and may be present in stable populations, but some characteristic of their population sizes or distributions make them conceivably at risk in the long term. **Indeterminate** applies to taxa known to be endangered, vulnerable or rare but for which there is insufficient information to say which category is the most appropriate. **Insufficiently known** species are those for which there is insufficient information on which to base a judgement concerning either their abundance and distribution, or the degree of threat they face.

We stressed above that conservation problems are best phrased and answered in terms of probabilities. The language of risk reflects this perspective. Mace and Lande (1991) suggested the IUCN categories should be assessed quantitatively and they proposed the following risk categories.

1. **Critical**: a 50% probability of extinction within 5 years or two generations, whichever is longer.

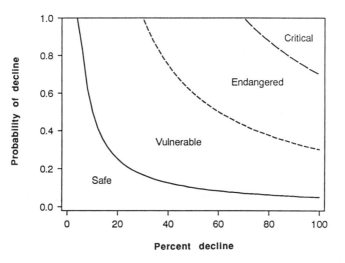

Figure 1.7 Levels of threat as a function of probability of decline and percentage decline (the proportion of the original population that is lost). Time is fixed at 50 years. (From Akçakaya, 1992; reproduced with permission of Elsevier, Amsterdam.)

2. **Endangered**: a 20% probability of extinction within 20 years or ten generations, whichever is longer.
3. **Vulnerable**: a 10% probability of extinction within 100 years.

Risk assessment in its most general form involves the probability of partial or total loss of a population, i.e. the risk of quasiextinction (Ginzburg *et al.*, 1982). Akçakaya (1992) recognized that the classification of risk involves three parameters, namely, time, probability of decline and percent decline. He generalized the approach taken by Mace and Lande (1991) to form a three-dimensional representation of conservation status. Figure 1.6 is a two-dimensional slice of this representation, with percent decline fixed at 100% (extinction).

Akçakaya (1992) calls these curves 'iso-risk' curves. The three points on Figure 1.6 represent the threat categories proposed by Mace and Lande (1991). Threat categories in Akçakaya's scheme are not represented by a single point but as areas on the figure. The time axis may represent years or generations equally well. The curves asymptotically approach 100% probability of extinction in infinite time, reflecting the fact that all populations eventually become extinct.

Alternatively, time may be fixed and percent decline and probability of decline may be displayed. In Figure 1.7, time is fixed at 50 years.

By tracing the iso-risk curves in this plane, it can be seen that the probability of decline of a population decreases with increasing percentages of decline. This says that there is a greater chance of partial loss than of total loss of any population. The shapes of these curves are hyperbolas, based on the conception that the risk (threat) of an event is the *product* of the probability of that event (say, 20%) and the loss that results from the event if it actually happens (a 50% decline). The threat is given by the product of these two quantities ($0.5 \times 0.2 = 0.1$). Note that the points at which the three curves in Figure 1.6 intersect the 50-year line correspond to the values of the iso-risk curves when they reach 100% decline in Figure 1.7.

The iso-risk curves in Figure 1.7 give constant products of the probability of decline and percent decline. For example, a population that has a 50% chance of declining by 10% in the next 50 years ($0.5 \times 0.1 = 0.05$) is at the same level of threat as a population that has a 5% chance of extinction within the same period ($0.05 \times 1 = 0.05$). This approach is particularly useful in evaluating different impacts such as: Does a 60% chance of a 60% decline pose more or less of a threat than a 30% chance of an 80% decline?

Defining categories is a much easier task than assigning species to them. Mace and Lande (1991) suggest some general rules based on population size that may be used for this purpose. Such rules are useful only when no other information is available. The models we outline in this book may be used to reach objective conclusions about the conservation status of species. Whereas biological intuition is essential for constructing models, it may be less useful for assigning species to risk categories. The functional relationships determining the dynamics of a population are mechanistic and it is possible for biologists to make dispassionate evaluations of them. However, people are poor judges of the risks that affect their day-to-day living, and we have no reason to believe they will be any better at judging risks for other species. The problem is a difficult one because often the autecological data and demographic studies necessary to place a population into a classification such as the one above are missing. In the absence of data and a quantitative analysis of risk, it is not possible to do anything except use value judgements and biological intuition.

1.2 CAUSES OF EXTINCTION

It may seem that extinctions in past times do not have much to do with model building and risk assessment. But an important source of risk information is experience, of which the fossil record is a part. It is not possible to await the outcome of direct observations on extant taxa. We do not have enough time and, in any case, our job is to prevent extinctions. We may look for symptoms of high levels of risk (such as small population sizes, or population decline through time) or we may examine the historical record.

The single most important lesson to be learned from the fossil record is

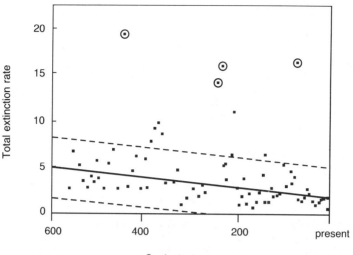

Figure 1.8 The total extinction rate of marine animal families over the last 600 million years. The extinction rate is measured as the number of families per million years (my). The circled points are sampled rates that are significantly higher than the remainder, and represent mass extinction events. The regression line and 95% confidence limits were set ignoring the outliers. The background extinction rate appears to have declined over the last 600 million years, which is consistent with the idea that optimization of fitness should increase with evolutionary time (after Raup and Sepkoski, 1982).

that all species eventually become extinct. More than 99% of the species that have ever lived on earth are now extinct (Simpson, 1952). There have been of the order of 10^8 extinctions in the past, of which a mere 200 000 have been found in the fossil record (Raup and Sepkoski, 1984). It is safe to assume that all species that currently exist will become extinct at some time in the future, with or without the intervention of humans. The only thing that differs among species is how long it will be before their demise (see Stanley, 1987; Nitecki, 1984).

When palaeontologists work through stratigraphic sequences identifying and counting animals and plants, they observe that existing taxa disappear and new taxa appear. If a taxon that was common at one time in a number of different locations is absent from strata representing later times, it is reasonable to presume that the taxon has become extinct (see Cronin and Schneider, 1990; Valentine, 1990). Using these observations, it is possible to quantify the number of taxa that become extinct between two points in time. Patterns of extinction (and speciation) vary through time and a great deal of interest has been generated in trying to explain them.

Palaeontologists often work with taxa other than species because of the

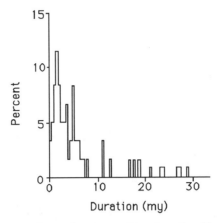

Figure 1.9 Histogram of species longevities for the Neogene radiation of planktonic foraminifera. Median longevity is 3.6 million years (my) and modal longevity is 2 my. (For 59 taxa; after Levinton and Ginzburg, 1984.)

difficulties of defining a species in the fossil record, and because the fossil record is so patchy it is difficult to know when a species becomes extinct. Often, they estimate extinction rates for higher taxa such as genera and families instead. The fossil record is far from a complete record of species that have gone before, and it is also biased. Fossils come predominantly from areas where sedimentary material accumulated (river deltas and lake beds). Animals and plants from other environments, and animals lacking large, heavy bones, are not well represented.

Despite these difficulties, studies of extinction patterns have led to a distinction between two qualitatively different rates of extinction. The first is termed the **background extinction rate**. It is the global average rate of extinction of taxa through time. The second is termed **mass extinction**, and refers to short periods in geological time in which many species in unrelated taxa become extinct. The magnitude, frequency and periodicity of these phenomena were investigated in two important papers by Raup and Sepkoski (1982, 1984; Figure 1.8).

The last mass extinction marked the end of the Cretaceous period about 65 million years ago. About two-thirds of all species then living became extinct, among them most of the extant dinosaur species. This time also marks the beginning of the adaptive radiation of mammals. There has been considerable debate in the palaeobiological literature concerning the causes of this mass extinction. Explanations range from supernovae and comets to loss of habitat, climatic change and random chance (for discussion and reviews, see Raup and Gould, 1974; Alvarez *et al.*, 1980; Levinton and Ginzburg, 1984; van Valen, 1984; Jablonski, 1986; Marshall, 1988).

Because of the biased and incomplete nature of the fossil record, it is difficult to be certain of the background rate of extinction in the remote past. As early as 1832, Charles Lyell estimated it to be about one species per year, a figure close to estimates made recently (van Valen, 1985). This figure represents the rate averaged over all species throughout the world. However, even if we ignore mass extinctions, the rate is highly variable among continents, and among taxa and at different times in the geological past. Many groups have undergone periods of unusually high extinction rates, after each of which adaptive radiation has rapidly increased global diversity (Stanley, 1985, 1987).

The average lifespan of a species is a few million years. However, there is a great deal of variation among the lifespans of taxa, even among closely related taxa (Levinton and Ginzburg, 1984; Gingerich, 1985; Stanley, 1985; Figure 1.9). Characteristics that seem to bestow resistance to extinction in 'normal' periods, such as ecological generalization and broad geographic range, do not appear to be particularly advantageous during periods of mass extinctions. Lineages and adaptations can be lost for reasons unrelated to their survival values during periods between mass extinction events (Jablonski, 1986).

Figure 1.9 shows that the distribution of extinctions for this taxon of planktonic foraminifera is skewed to the right. In fact, it looks much like the statistical distributions for the expected time to extinction predicted by most models of individual taxa.

1.2.1 Extinction in recent times

It is difficult to estimate how many species there are on earth. About 1.4 million have been documented by science. The true number is likely to be between 10 and 50 million (see the reviews by May, 1988, 1990; Gaston, 1991; Erwin, 1991).

In the last few decades, concern has mounted that many species may be lost within the next few decades. There are a number of phenomena contributing to the loss, including the worldwide rate of habitat destruction, the rate of destruction of species rich rainforest for fuel, timber and agriculture (Myers, 1981, 1988; Mares, 1986; Nectoux and Kuroda, 1989; Given, 1990; Collins et al., 1991), the potential for change in the global environment caused by the build-up of carbon dioxide and other gases in the earth's atmosphere (Dobson et al., 1989) and observations on the accelerating rate of observed extinctions in the recent past (Ehrlich, 1986; Stanley, 1987).

Remarkably little is known about the process of global extinction, partly because it is difficult to determine whether or not a species is extinct, or merely hard to find. It is far easier to look at local extinctions. Terborgh (1974; Terborgh and Winter, 1980) worked with data on the chances of local extinction of populations of birds and Diamond (1984) and Pimm et al. (1989) extended this work. They looked for characteristics of species that were more likely to become extinct locally ('extinction prone' species) and found the main predictor of local extinction of birds on British islands is population size. Of course, there is no simple extension of these results based on populations to the level of species. We cannot assume that species that are prone to local extinction are prone to global extinction. We shall explore this topic further in Chapter 5 where we provide an approach for modelling metapopulations.

Extinctions on two islands

The islands of Madagascar and Australia have rich, endemic floras and faunas. Both support relatively small human populations that have, over the last few hundred years, cleared or severely disturbed the majority of their natural environments.

Jenkins (1987) reviewed what is known about changes in the abundance and distribution of native plants and animals on Madagascar. More than 80% of the 8000 floral species are endemic, as are 106 of the 197 species of resident birds and more than 90% of the 400 species of amphibians and reptiles. A total of 7 species of carnivores, 10 rodents and all of the world's lemurs are indigenous.

Although the human population is not large (just over 10 million in 1985), more than 85% live in rural villages of less than 2000 people. At least 70% of the island no longer has significant woody plant cover and the principal agent for clearing is slash-and-burn cultivation. The land can sustain a maximum of 10 to 15 rotations before deterioration of soil structure and nutrient content make further cultivation impossible. Jenkins (1987) made a conservative estimate of the rate of conversion of forest to uncultivable land of 10 000 ha per annum.

The loss of habitat to agriculture has direct consequences for the natural populations of plants and animals. Of the birds, one species is presumed extinct and 27 others are threatened. Of the 28 primate species, 24 are threatened, of which 5 are endangered. A total of 22 amphibians and 70 reptiles are threatened or are being considered for inclusion in an IUCN category. However, only one reptile, a tortoise from Cape Sada, is known to be critically threatened. There is insufficient information to assign most species to one or another of the IUCN categories. For example, the colubrid snake *Liophidium apperti* is known from a single collection made in 1968 in deciduous forest near Befandriana-sud. The forest is now cleared but for a few isolated trees. Such patterns of richness, endemism and lack of critical information were repeated in each of the taxonomic groups that Jenkins (1987) reviewed.

Like Madagascar, the island of Australia supports a rich, highly endemic biota. Land-use changes since Europeans first settled about 200 years ago have included complete removal of vegetation in urban and mining developments, replacement of existing ecosystems with agriculture and exotic forestry plantations and modification of ecosystems with pastoralism and timber harvesting (Hobbs and Hopkins, 1990). Despite the relatively small, highly urbanized population of less than 20 million people, most of the land area has been partially or completely disturbed and the process continues. For example, 54% of the land cleared for agriculture in the south-west of Western Australia was cleared between 1945 and 1982 (Saunders *et al.*, 1985 in Hobbs and Hopkins, 1990).

Burbidge and McKenzie (1989) and Recher and Lim (1990) reviewed the changes in the distribution and abundance of the vertebrate fauna of Australia over the last 200 years. Excluding New Guinea, which has its own rich and unique fauna and with which Australia shares many species, more than 90% of the 1600 terrestrial vertebrate species are endemic. Of these, nearly 300 are considered endangered. A total of 17 mammals, 3 birds and 1 lizard are known to have become extinct. Many other species have undergone dramatic reductions in range and abundance. For example, 8 mammal species that were once present on the mainland persist only on small off-shore islands.

Extinction rates and declines in abundance and range have been highest in regions where settlement first occurred, attributable to land use, habitat loss and fragmentation, over-exploitation, and the spread of exotic herbivores, predators and disease. Recher and Lim (1990) note that dramatic declines leading to extinctions of the fauna are measured in decades, not centuries, and they predict accelerated losses among birds, reptiles and amphibians in the near future.

Comparing recent extinction rates with those in the past can only be done for birds and mammals, where there is reasonably complete knowledge of both the extant taxa and the fossil record. In 1986, Ehrlich used data from these two taxonomic groups and estimated that extinction rates at that time were 4 to 40 times higher than those than have persisted over the last 50 million years. He predicted that they soon would be about 100 times the background level. If current land-use practices continue, about half of all species now living will be lost within the next 50 to 100 years (Myers, 1981; Simberloff, 1986b; May, 1988, 1990; Wilson, 1988; Given, 1990). The twentieth and twenty-first centuries will easily qualify as a mass extinction of geological proportions, rivaling the event of 65 million years ago.

1.2.2 Habitat loss

The thing that is different about the mass extinction event currently under way is that we know its ultimate cause and we are in a position to do something about it. By using the information in the fossil record and observations on the loss of extant species, it is possible to characterize the particular factors that cause species to become extinct. The role of quantitative models for risk assessment is to evaluate the impact of these different factors.

The ultimate cause of most current extinctions is the size of the human population (Ehrlich and Ehrlich, 1990). The number of people on earth does not itself result in species extinctions. The problem is that people need space in which to live, space that was used in other ways by other species. For example, the human populations of Australia and Madagascar clearly contribute to the decline in range and abundance of native species directly through land use, and indirectly through the introduction of feral animals, plants and diseases.

There is a fundamental relationship between the size of a population and its chances of extinction because all populations are engaged in a random walk, some more random than others. On a geological time scale, large populations have an adaptive advantage as well. Fisher (1930) pointed out that relative abundance guarantees an evolutionary advantage because abundant species will make the most rapid evolutionary progress. He postulated that species diversity is a balance between extinction of rare, slowly evolving species, and speciation in abundant, rapidly evolving taxa.

When habitat is lost to a species, the total abundance of all factors necessary for survival and reproduction (such as food, living space, nesting sites) is reduced. If there are fewer of these resources, the environment will support fewer individuals of the species. The habitat that remains frequently is fragmented into more or less discrete patches interspersed with a suboptimal or hostile environment. Habitat loss results in smaller populations which in turn results in increased extinction risks. On a global scale, the more habitat that is lost, the more species will become extinct in any given period of time.

For a species, habitat loss does not evoke an adaptive response. Genotypes that can extend their environmental tolerance to use other resources would have an advantage over members of the same species. However, unless the species is pre-adapted to the modified environment and can occupy sites within it, the countervailing pressures of competition and predation will preclude the extension of environmental tolerance within the remaining habitat, with the result that extinction may be random with respect to a species' adaptations (Gould and Eldredge, 1977; Futuyma, 1986). Habitat loss may be the single most important cause of extinctions, both in the geological past (Gould and Eldredge, 1977) and at present (May, 1988, 1990).

The relationship between the factors that limit population size and distribution and the chances of extinction is not a simple one because different species are rare, restricted or scarce for different reasons.

There are doubtless many cases where species exist in small numbers within a small geographic range, yet persist in a stable population for many millions of years. It is conceivable that some of the factors that lead to the relative rarity or restricted geographic range of species may be independent of those that cause extinction. Some species have limited environmental tolerance (Brown, 1984) or are tolerant of a wide range of environmental conditions but nevertheless lack available habitat (Bengtsson *et al.*, 1988; Burgman, 1989; Prober and Austin, 1991).

Threats to the persistence of the spotted owl
The spotted owl (*Strix occidentalis*) in the United States is a good example of a recent, well-documented decline in a species that is almost certainly caused by the dwindling amount of its available habitat. The owl depends on forest older than about 250 years. Each nesting pair may require as much as 2 to 8 km^2 of forest. The area of suitable habitat in the Pacific north-west fell from 11 040 km^2 in 1961 to 2190 km^2 in 1984. The USDA Forest Service (see Simberloff, 1987) estimated that the remaining 2500 pairs were declining towards extinction. While the current population may be more or less replacing itself, models indicate rapid extinction if current and projected logging programmes are maintained because there would remain too little habitat to support viable populations. For discussion of the models, their results and economic and social implications, see Dixon and Juelson (1987), Simberloff (1987), Lande (1988a), Carey *et al.* (1990) and Noon and Biles (1990).

The correlation between population size and risk of extinction will be weakened if, for example, a large population were struck by a catastrophe. Catastrophes are rare and unexpected events, environmental changes of great magnitude and suddenness. They cannot be countered by evolutionary change or by the presence of a resistant genotype. It is indisputable that catastrophes such as widespread volcanic eruptions and large meteorites have occurred in the earth's past (Alvarez *et al.*, 1980). It is less clear how important these events have been in eliminating large numbers of species, more or less simultaneously (see van Valen, 1984, for a review).

These observations are hardly surprising. There must be a certain amount of bad luck involved in the extinction of any species. That is to say, extinction is a random process, and given a large enough sample size (millions of species and millions of years) anything is possible. The possibility of sheer bad luck in competing with other species for available habitat, combined with competitive exclusion, has been the basis for the formulation of various lottery competitive models (Sale, 1977; Chesson and Warner, 1981). There is ample empirical evidence to support the notion that historical accident has indeed been important in determining the current distribution and relative abundances of many species (see, for example, Gilbert and Lee, 1980; Cody, 1986; Burgman, 1989; Bengtsson *et al.*, 1988).

The human population will continue to increase throughout the foreseeable future. The added risks faced by species through loss of habitat will continue to mount as a result. The role of risk assessment is to identify those specific factors that impinge most critically on the chances of extinction of a population. With it, we can identify the most efficient ways to reduce the risks with the resources at our disposal. We can inform society of the real risks faced by species so that objective decisions can be made concerning the trade-offs required to improve the chances of persistence of species.

1.2.3   Competition and environmental variability

Habitat loss is not the only reason that species become extinct. There are numerous other factors that may affect chances of extinction, all of which may be accounted for in models of a species dynamics. Factors that may lead to global extinction were listed by Soulé (1983). Apart from habitat loss he included rarity, genetic effects, environmental change, catastrophes, predation, competition, demographic accidents and social dysfunction at low densities. Humans may be responsible for many of these factors; a few of them would probably have occurred had humans been present or not.

Factors contrast with habitat loss because they may elicit an adaptive response from the population (Simpson, 1953; Futuyma, 1986, Ch. 12). When novel conditions are experienced by a species, the frequencies of existing genes may change in response. Successful phenotypes in the new regime are, by definition, those that survive and have more offspring than other phenotypes. If genetic variation is present in the population, the genes that give rise to more successful phenotypes will tend to become more frequent.

Examples of the effects of competition involving alien, invasive species are relatively common in the literature. Some documented cases have led to local or global extinctions, while others have led to reductions in total numbers of individuals. However, without the benefit of manipulative experiments, it is difficult to prove that competition is responsible for changes in any one case. For example, the red squirrel (*Sciurus vulgaris*) is being replaced in Europe by the introduced grey squirrel (*Sciurus carolinensis*), although the mechanism for replacement could involve competition, environmental change or disease (Reynolds, 1985; Kenward and Holm, 1989).

Competition among endemic species typically has less spectacular results than the effects of alien species probably because most of the action occurred before we started looking for it. The species distributions we observe today are the result of past competitive interactions (Connell, 1983, cf. Ferson *et al.*, 1986). As a result, it is more difficult to document the resulting decline of population abundances. In one example, Mehrhoff (1989) suggests that increased competition for resources from surrounding vegetation may be responsible for the decline of a population of the endangered orchid, *Isotria medeoloides*.

Predation, disease and pollution all may threaten the persistance of species and recent examples of extinctions are quite common. For example, the predatory mongoose (*Herpestes auropunctatus*) has eliminated several bird species on Caribbean islands (Simpson, 1953). Overharvesting by humans may best be seen as a special case of predation and there are many examples of species driven to extinction or near extinction, including passenger pigeons and buffalos in the United States, bison in Euope, and white and black rhinoceros in Africa.

Specific environmental parameters such as rainfall and temperature have been shown in numerous studies to impinge directly on the reproduction and survival of individuals in plant and animal populations (see Mehrhoff, 1989, for orchids; Parsons, 1989, for *Drosophila*). Unusual conditions can threaten the persistence of populations. Drought has caused the local extinction of checkerspot butterfly (*Euphydryas*) populations in the western United States (Ehrlich *et al.*, 1980). Similarly, extended drought is implicated in reduced fecundity of white rhinoceros (*Ceratotherium simum simum*) in Africa (Conway and Goodman, 1989).

The period over which environmental variation occurs is important and the impact of environmental change will depend on the longevity of a species. That is, the life strategy of a species is related to its susceptibility to environmental extremes (Rabinowitz, 1978). For example, very long-lived species must survive short-term fluctuations in the environment and reproduce when conditions are suitable.

We noted above that we are concerned in this book mainly with the dynamics of individual populations and species. When seen from the perspective of community dynamics, environmental variation may reduce the chances of extinction of species. Chesson and Warner (1981) point out that environmental variability may promote coexistence, allowing the persistence of some species in habitats from which they would otherwise be excluded by competition.

Extinction of the bald ibis in Turkey
The causes of extinction of species, either locally or globally, may be complex and involve several causes. The bald ibis, *Geronticus eremita*, is a globally threatened species that once inhabited central Europe, north Africa and the Middle East. It is an unusual ibis

because it lives in arid and semi-arid regions and nests on cliffs or rocky ledges. It disappeared from central Europe in the mid-seventeenth century probably because of climatic changes and hunting. Until recently it survived in two separate populations, one in north-west Africa and one in east Africa and Turkey. There are important morphological and ecological differences between these two populations. For instance, the north-west African population is resident whereas the Turkish population is migratory.

Akçakaya (1990) documented the decline and ultimate loss of the Turkish population. Between the early 1900s and 1953, census information suggests the population was more or less stable between 1000 and 1300 birds. It declined dramatically over the next 20 years, the main reason for which was the extensive use of pesticides, particularly DDT, against locusts and malarial mosquitos near the nesting grounds at Birecik. Concentrations of pesticides were so high that people fell ill, domestic animals died and indigenous species in the area were largely destroyed. About 70% of the ibis population died from acute insecticide poisoning. Human disturbance contributed to the problem and almost no young were produced between 1953 and 1972, when the population had fallen to about 20 breeding pairs.

A conservation project was initiated by the WWF in 1973 and was implemented by the Turkish Government. However, the captive breeding and reintroduction programme failed because reintroduced birds failed to join the wild population in migration. Complex learning patterns associated with both feeding and migration had been disrupted in captivity. In 1989, only three birds returned to Turkey from east Africa, two of which died subsequently in a storm.

There are many factors that led to the demise of this population. They include habitat loss, long-term weather changes, hunting, human disturbance of nesting sites, pesticide poisoning, disruption of social behaviour, and short-term environmental extremes.

1.2.4 The genetics of species at risk

There are two aspects of genetics to be considered in relation to conservation; the conservation of genetic variation, and the relationship between genetic variation and the vital rates (fecundity and survivorship) of individuals. The latter relationship is the key to using genetic information in models for demographic risk assessment.

Wild populations may have existed in sizes close to equilibrium abundance for many generations. When populations are reduced in size over a short time period (by harvesting, loss of habitat, or some other factor), or when populations that were previously separate are joined, the effects may be deleterious. Small populations are susceptible to genetic deterioration through genetic drift which may lead to the loss of genetic variability and the expression of deleterious, recessive alleles. The result is termed inbreeding depression. Frequently, the result is reduced fertility in adults and increased juvenile mortality (Ralls and Ballou, 1982a,b; see reviews by Simberloff, 1988 and Burgman et al., 1988).

The processes that led to these reductions in survivorship and fecundity have been characterized. They include inbreeding depression, outbreeding depression, loss of fitness following hybridization between populations and the expression of deleterious alleles (see Soulé, 1980, and Simberloff, 1988). These processes have been documented in both wild and captive populations. Ralls et al. (1983) noted them in Californian sea otters, O'Brien et al. (1986) relate reduced fitness in the cheetah to loss of genetic variation and Wiens et al. (1989) link the decline of the shrub *Dedeckera eurekensis* to its genetic status. Other examples are provided by Bonnell and Selander (1974), Greig (1979) and Ralls and Ballou (1982a,b).

Genetic variation in the European bison
One of the earliest studies of the genetics of a threatened species was conducted by Slatis (1960). The European bison, *Bison bonasus bonasus* was hunted almost to extinction by the mid-nineteenth century. After an analysis of genealogies of the remaining herd of about 200 individuals, Slatis found a positive relationship between the degree of inbreeding and the frequency of perinatal and juvenile death. He ascribed the deleterious effects of inbreeding to the presence of recessive lethal genes (an average of six were present in each outbred ancestor), rather than to a loss of generalized heterozygosity. The results imply rapid selection against recessive lethals that would otherwise have persisted in the population indefinitely.

1.2.5 The development of methods for risk assessment

The earliest attempts to predict extinctions specifically for problems in conservation and wildlife management were based on the equilibrium theory of biogeography formulated by MacArthur and Wilson (1967). The theory suggests that the number of species on an island is a dynamic balance between speciation, immigration and extinction. Extinction rates are roughly inversely correlated with population sizes and population sizes are positively correlated with area. Immigration rates are negatively correlated with distances to other islands (Figure 1.10).

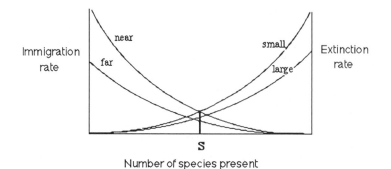

Immigration rate

near

far

small

large

Extinction rate

S

Number of species present

Figure 1.10 Equilibrium model for the biota of several islands of varying distances from a source area, such as a continental land mass, and of varying size. The vertical line and the bold faced **S** give the number of species expected on a small island close to the source of species. (After MacArthur and Wilson, 1967, p. 22.)

Thus, for small islands and for islands a long way from a source of immigrants, the number of species will be fewer, a result of the dynamic equilibrium. An increase in distance makes it harder for immigrants to reach an island, lowering the immigration curve (Figure 1.10). An increase in area means most species will be present in larger populations, lowering the extinction curve. the extinction curves increase from left to right because, on an island of a given size, if there are more species present, each species will consist of fewer numbers and extinction will be more likely. At equilibrium, a balance is reached where the number of new immigrants per unit of time equals the number of extinctions. The immigration rate is determined by the distance of the island from the mainland and the extinction rate is determined by the size of the island.

The principles of island biogeography were used to answer questions such as: 'What is the best way to partition an area of land to maximize the number of species contained in a reserve system?' The idea gained wide acceptance in the 1970s and was used to formulate general rules for the design of wildlife refuges (Wilson and Willis, 1975; Diamond, 1975; see Simberloff, 1986a).

A number of reviews have expressed doubt concerning the validity and usefulness of the theory (Margules and Usher, 1981; Soulé and Simberloff, 1986; Zimmerman and Bierregaard, 1986; Burgman *et al.*, 1988). The theory is phrased in terms of the equilibrium number of species. It considers only two factors (size of the island and its distance from the mainland) in determining this variable, and ignores many others (such as the demographic properties of species and the geographic configuration of islands). Thus the theory of island biogeography may not be able to predict the equilibrium number of species in a system of habitat patches with accuracy. The theory was fraught with problems, including a lack of unequivocal empirical support and a lack of unique, precise or reliable predictions (Boecklen and Gotelli, 1984; Simberloff, 1986a; Burgman *et al.*, 1988). One of the most important inadequacies in using it for wildlife conservation is that the theory tells us nothing about which species are likely to become extinct and it ignores whatever autecological information is available.

Nevertheless, island biogeography has been a very useful model in conservation biology, especially as a conceptual framework for focusing on relationships among area, distance and species number. In Chapter 5, we shall consider it further where it is treated as a special case of a more general class of models for spatially structured populations.

While the ideas for island biogeography were being formed, a parallel set of ideas were developing more or less independently, concerned with the risks faced by individual populations. In 1949, Allee *et al.* defined a minimum viable population size as that below which populations will increase their chances of extinction. In a paper published in 1962, Moore outlined the consequences of habitat destruction and fragmentation: increased risk of global extinction through failure to disperse between local populations, edge effects, inbreeding depression and accident. He defined the smallest viable size of habitat as the smallest which supports a viable population of its own key species. Key species are those that, when removed, precipitate a radical effect on the whole habitat or community. MacArthur and Wilson (1967) and Simberloff and Abele (1976) both recognized that there may exist a critical threshold, a minimum area supporting a minimum population size below which species will rapidly become extinct. The relationship between population size and the chances of extinction of a population is implicit in the dynamic equilibrium theory of island biogeography.

One of the most important developments in conservation biology has been the acceptance of the notion of risk. Shaffer (1981) defined the minimum viable population arbitrarily as the smallest isolated population having a 99% chance of surviving for 1000 years. This definition embodies the concept that there exists an initial threshold size below which populations face unacceptable risk. For example, assume we agree with Shaffer (1981) that it is acceptable that a given species faces a 1% chance of extinction within the next 1000 years, but if the risk is any higher the situation is unacceptable. To determine that this risk corresponds to an initial population size of, say, 2000 individuals, we shall need to model the dynamics of this species, its environment and all the factors that will tend to push it towards extinction.

Gilpin and Soulé (1986) termed the comprehensive analysis of all the factors that may cause a species to go extinct, population viability analysis (see Shaffer, 1990). The relationship between a minimum viable population and practical conservation is that a population of adequate size will require a certain area of land, although it is by no means easy to measure how much land is necessary to support a given number of any species (Gilpin and Soulé, 1986).

The evaluation of risks requires quantitative analysis of population dynamics. While conservation biologists were busy defining the role of risk assessment in wildlife management, the groundwork to develop the necessary mathematical tools had already been completed. The notion that stochasticity plays an important role in determining the expected lifetime of a species has long been of interest to mathematical demographers. The birth-and-death model, which we shall examine in more detail later, treats the effect of demographic variability on population persistence and was developed by Feller (1939) and later by Kendall (1948) and Bartlett (1960).

Work relating the effect of environmental stochasticity to population persistance was begun by Lewontin and Cohen (1969) on exponential growth in discrete time with random growth rates, and by Levins (1969) on the logistic equation with random growth rates and random carrying capacities in continuous time. One of the earliest models of a natural population addressing the question of species decline in a variable environment was formulated by Lebreton (1978). He accounted for both age structure and density dependence. The paper in which he presented his work was published in a statistical journal and went largely unnoticed by conservation biologists.

Just as many of the methods for estimating the effects of demographic and environmental uncertainty were developed before conservation biologists needed them, so the methods for dealing with genetic uncertainty were developed much earlier. The impetus for the development of methods for quantitative genetics came from agriculture, animal and plant breeding, and evolutionary theory. The relationship between population size and life history parameters such as reproduction and survival has been a central question in genetics since the early 1900s (see Wright, 1977). Genetic problems including the expression of deleterious recessive alleles and the loss of heterozygosity were considered in the context of conservation of natural populations by Slatis (1960), Moore (1962) and Berry (1971) long before an integrated approach to conservation biology was developed.

The methods for risk assessment were developed originally in quantitative ecology, public health and safety (Ginzburg et al., 1982), mathematical demography and quantitative genetics. Conservation biology is fortunate to have had many of its fundamental methods ready made. As a result, the development of conservation biology has been relatively rapid.

1.3 SUMMARY

All populations live in uncertain environments and their long-term survival depends on the short-term, probabilistic fluctuations in their population abundances. Stochastic models allow us to evaluate the risks faced by natural populations, subject to the limitations of the model.

The changes in abundance in any population through time may be characterized, at least in part, as a random walk. Data are scarce for most natural populations, even though our understanding of the important deterministic and stochastic processes may be good. A stochastic model is just a deterministic model in which one of the parameters varies randomly. Risk is the chance of an adverse advent. Risks for natural populations are evaluated by estimating the probabilities of these events using stochastic models.

Quantitative estimation of risk is necessary because people are poor judges of risk levels. The application of risk analysis to the demography of natural populations requires the application of new terms and different ways of assessing impacts on populations. The best approach is to quantify the probability of extinction within a given period of time. The estimated time to extinction of a population is less informative because the statistical distribution of time to extinction is skewed to the right, so the average time is a poor representation of what is most likely to happen to a population.

The sources of uncertainty that contribute to changes in population abundance can be classified as phenotypic, demographic, environmental and spatial variation. The terminology for population size and distribution (abundance, commonness, rarity, range, habitat specificity, scarcity) describes different properties of a population than the terminology for risk (threat, vulnerability). The categories for degrees of risk may be best applied to populations after quantitative evaluation of risk levels.

There is ample historical evidence for dramatic changes in species abundances and species extinctions. Recent extinction rates rival the mass extinction event that occurred 65 million years ago. The mass extinction event currently under way is a result of the size of the human population. The human population makes itself felt largely through the destruction of habitat of other species. The consequent decrease in natural population sizes adds to the other factors that tend to drive species to extinction, such as competition, predation, disease, extreme environmental conditions and the deleterious effects of inbreeding in small populations. Risk assessment is essential if we are to allocate scarce resources to conservation and wildlife management as efficiently as possible, thereby minimizing the number of species that will become extinct within the next few years.

2 White rhinoceros on Ndumu Reserve

To introduce the concepts we talked about in Chapter 1 in a more concrete fashion, we shall develop a stochastic model for a population of white rhinoceros (*Ceratotherium simum simum*). These particular animals live on the Ndumu Game Reserve in Africa and their demography is described by Conway and Goodman (1989). Ndumu is a small game reserve of about 116km^2 on the Mozambique coastal plain. The population of white rhinoceros was established with the introduction of 20 individuals between 1961 and 1963. Currently, the population is increasing in size. Our aim is to build a model for the change in size of this population, but before we begin to write equations we shall introduce some general concepts that will be useful for this model and later on.

The complexity of a model must be a compromise between reality, available data and the use to which the model is put. When data concerning a population are scarce, any model of it must by necessity be simple. An unstructured, or single-dimensional, model is one in which such things as the proportion of individuals in different ages or life history stages, or in different spatial patches, are ignored. Population size is represented as a single variable, hence the terms unstructured and single-dimensional. Survival from one year to the next or the chances of producing viable offspring may be closely related to the age, developmental stage or location of an individual. When unstructured models are used, it does not mean that these factors are unimportant. Rather, it means that their effects are accounted for by a set of assumptions. If the assumptions are valid, the model will be adequate.

No model can replace the acquisition of data. The objective of the model is to improve our understanding of the population, to test our ideas and acquire more data with the ultimate aim of gaining sufficient knowledge to preserve the species. Thus, when data and understanding are in short supply, it is advisable to begin with a model that summarizes what we know. The predictions it makes should be treated with caution. Very simple models of population dynamics can be constructed that ignore considerations of the ages, life history or developmental stages, and spatial structure of species. The barest details of population distributions or individual behaviour may provide enough information to explain, in general terms, how population size is regulated. Usually, simple models for demographic risk assessment will be most useful to help us understand generalities such as the effect of environmental and demographic variability on population growth rates, or the effect of harvesting on expected population size. One of the most important uses of models is that they focus attention on important items of data that are missing, or on items of data that need to be estimated with much greater precision. This focus is related to sensitivity analysis, which we shall address in Chapter 3.

It is worth noting that simple models may in fact require as much information as models that account for details of population structure. If there are fewer parameters in a model, there are more assumptions. For example, if we ignore the different ages of individuals, we assume age plays no important part in the average chances of an individual dying or reproducing. If we are to judge the validity of that assumption, we would need information on the relationship between age and vital rates.

As a result, conventional wisdom dictates that we should place little confidence in the accuracy of quantitative predictions made from simple, general models. These models often proceed from assumptions that are greatly oversimplified and are usually employed to generate qualitative answers to very general questions. However, if a model diverges from the details of the real world, it does not necessarily follow that it will be a poor predictor of ensemble properties such as the average population size at some time in the future, or the chances of crossing a given population size threshold. The values we assign to the parameters, and their relationships to one another in the model, embody all assumptions. We do not suggest that the kinds of models described below are sufficient for every case. Rather, when time is short and data are critically scarce, they may be the only methods. Conservation biology often deals with cases where time and resources are critically scarce. When urgent decisions are required and you have to decide one way or the other, it may be better to use a crude model than to use intuition alone.

2.1 FORMULATING A BIRTH-AND-DEATH MODEL

The total number of individuals (N) in a fixed region of space can only change because of births, deaths, immigration and emigration. Change in population size over a discrete interval of time from t to $t + 1$ can be described by the equation

$$N_{t+1} = N_t + B - D + I - E \tag{2.1}$$

where B and D are the total number of births and deaths respectively during the time interval from t to $t + 1$, while I and E are the total number of individuals entering and leaving the region during the same time interval. Of course, we may replace immigration and emigration by processes that are mediated by humans, such as introductions, harvesting or poaching. Change in population size from t to $t + 1$ is given by $N_{t+1} - N_t$.

For the moment, let us assume there is no immigration to, or emigration from, the rhinoceros population. We can rewrite Equation (2.1) in terms of the difference between the average per capita rate of reproduction, or birth rate, (b) and the average per capita death rate, (d). The two parameters, b and d, can also be considered as the probability of having offspring and the probability of dying between one time and the next. We can calculate, in continuous time, the average rate of change in population size:

$$\frac{dN}{dt} = (b - d)N \tag{2.2}$$

where, by definition, the time interval, dt, had been reduced to become infinitely small. Thus, Equation (2.2) describes the instantaneous rate of population size change. This equation can be solved by integration to give the familiar equation for exponential population growth. At some time, t, in the future, given the initial population size, N_0, the birth rate and the death rate, the size of the population will be

$$N_t = N_0\, e^{(b-d)t} \tag{2.3}$$

where 'e' is a constant representing the base of natural logarithms. This equation may be written as

$$N_t = N_0\, e^{rt}. \tag{2.4}$$

The difference between the birth and death rates, represented by the term r, is called the 'intrinsic rate of natural increase'. When the birth rate is greater than the death rate, the population will grow exponentially, irrespective of its initial size (unless, of course, the initial population size is zero).

We may replace the term e^r with the term λ (lambda). Alternatively, we let r be equal to the natural log of λ. λ is called the 'finite rate of increase' of a population and it is also called R. It represents the proportion by which the population changes at every time step. For example, if $\lambda = 1.1$, then the population increases by 10% every time step; if it is 0.5, the population decreases by half.

We can use this opportunity to introduce some more useful terminology. A **dependent variable** (or **state variable**) is the thing in the model you want to estimate (such as population size). It is the entity which 'depends' on the other factors. These other factors are called **independent variables**. A model is an equation describing the relationship between the independent variables and the dependent ones. The operations encapsulated in the equation describe they way you think the state variable is modified by the independent variables. **Parameters** are those components of a model that mediate the relationship between independent and dependent variables.

Equation (2.3) above constitutes a model, a mathematical expression that allows us to estimate the population size at any time in the future. It has one dependent variable, N, one independent variable, t, and two parameters, b and d. N_0 is an initial condition, the initial value of the dependent variable: it represents the size of the population at time zero, usually the present.

It is always a good idea to look at the qualitative characteristics of a model before you estimate parameters and make predictions. For example, Equation (2.4) predicts a very strong relationship between the expected population size and the difference between the per capita birth and death rates. We can tell this by inspection: the difference between the vital rates is an exponent (it is raised to the power 'e').

2.1.1   Model assumptions

Whenever a model is constructed, it employs a set of assumptions reducing the complexity of the real world to manageable proportions. Assumptions are all those things not dealt with explicitly in the model but which must nevertheless be true for the model to provide reasonably accurate predictions. The model above makes a number of them.

It is obviously a vast oversimplification. There is no attempt to model the dynamics of populations comprehensively. The probability of dying, d, and the probability of having an offspring, b, are not likely to be the same for different individuals in a population. They will surely depend to some extent on the age of the individuals. Furthermore, these rates are not likely to be constant over time. Nor are births and deaths likely to be independent events.

A list of all the assumptions, at least those of which we are aware, is:

1. Populations will grow or decline exponentially for an indefinite period. This implies that population density remains low enough for there to be no competition among members for limiting resources. There are no density-dependent effects.
2. Births and deaths are mutually independent.
3. Births and deaths are independent of the ages of individuals. In real populations, births, deaths and the propensity to immigrate or emigrate are age, sex and density dependent. However, it turns out that even if birth and death rates are age dependent, the mean rate per individual will remain constant if the proportions of the population in each age class remain constant over time (see Chapter 4). It is sufficient that, on average, the constant parameters apply equally and independently to all members.
4. Birth and death rates (and immigration and emigration rates) are constant in time. This assumption will be violated by any external influence such as an epidemic, or by genetic changes in the population, or by a change in the sex ratio (the relative numbers of males and females). Often, models of populations deal only with females, assuming there are sufficient males present for reproduction not to be limited by a lack of them.
5. There is no variability in model parameters due to the vagaries of the environment or the population's demography. The model for

exponential population growth above (Equations (2.2) and (2.4)) is clearly a deterministic model: there is no uncertainty in its prediction. It says that at some time, t, in the future, the population size will be N_t, and it can be calculated exactly by the right-hand side of the expression.

Obviously, we need to know quite a lot about a population to be able to make all of these assumptions.

2.2   PARAMETERS AND INITIAL CONDITIONS

White rhinoceros on Ndumu Game Reserve are difficult to observe because they inhabit dense woodland and thicket. As a result, demographic data are scarce. Conway and Goodman (1989) report that there are 57 white rhinoceros in the population. The mean density of the population is $0.49\,\mathrm{km}^{-2}$, somewhat lower than densities of this species observed elsewhere in Africa. In areas of preferred habitat within the reserve, densities reach as high as $2.68\,\mathrm{km}^{-2}$. The population consisted, in 1986, of 45 adults and subadults (18 males and 27 females), 4 yearlings and 8 juveniles.

It does not really matter if we build a model for both males and females, or for females alone. It is a realistic assumption that there will always be enough males to go around. Females are usually the limiting sex in reproduction. It is easier to base the model on females, ignoring the males in the population completely. Conway and Goodman estimated the per capita birth rate (the chance each female has of giving birth to a female calf that survives to the next census) at 0.14 per year. The per capita death rate, the chance each female has of dying, is 0.08 per year. Thus, the maximum rate of increase (r) we might expect, given by the difference between birth and death rates, is 0.06. In 1986, there were a total of 35 females present on the reserve.

Conway and Goodman (1989) observed a 50% death rate in the transition from juvenile to yearling. Clearly, death rates are not the same at different ages in this population. If any other age class suffered nearly as high a death rate, there would very soon be no rhinoceros left on the Reserve. However, it is not strictly important that death rates are different in different age classes if the proportion of the population within each age class remains more or less constant (see Chapter 4). We shall assume this is so.

For the purposes of developing management plans, let us assume the period of interest is the next 50 years. Our starting date will be 1986, the last year of reported observations. Our object will be to look at the predictions of the model as it is developed so far, and to develop it further to include features that make it more realistic.

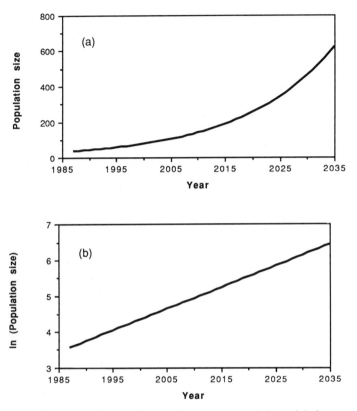

Figure 2.1 Deterministic prediction of an exponential model for growth of a white rhinoceros population: (a) population size is the number of females; (b) population size plotted on a log scale.

2.3   THE DETERMINISTIC PREDICTION

As a first step, we could evaluate the predictions of the deterministic version of our model. In this case, we have an analytical solution for N_t, given N_0 and r. We substitute the parameter values reported by Conway and Goodman in Equation (2.4), giving

$$N_t = 35 \ e^{0.06t}, \tag{2.5}$$

and solve the equation for each time step (Figure 2.1(a)).

The first thing to note is that the prediction is exact: there are no error bars or confidence limits. Second, if the population is allowed to grow for the next 50 years and there are no limits to the amount of habitat or other resources available, we could expect in excess of 650 females and a total population in excess of 1300 animals. When population size is plotted in a log scale (Figure 2.1(b)), growth appears linear. This transformation is important if proportional changes in population size are more important than additive changes. Any of the predictions for population size in this chapter and the chapters that follow could be represented in a log scale.

These results are not particularly satisfying. A population of 1300 would result in densities of 13 animals per square kilometre, about twice the density observed in more optimal habitat elsewhere in Africa (Conway and Goodman, 1989). This density is clearly unsupportable in the Park environment. It is very unlikely that the population would continue to grow in an exponential fashion throughout the next 50 years. Resources on the reserve would become depleted as density increased and survivorship and fecundity rates would fall as a result. One of the important assumptions of our model, that birth and death rates are constant from year to year, would surely become invalid.

Another troubling feature of the model is that in calculating the exponential expectation above (Figure 2.1), we used real numbers. For example, the prediction for the population size in 1987 is 37.2 rhinoceros and in 1988 it is 39.5. Clearly, we cannot in reality have 0.2 of a rhinoceros. One element of realism we can add to our model is to use integer numbers. We shall address this problem first.

2.4   ADDING DEMOGRAPHIC STOCHASTICITY

Above, we noted that the mean population birth rate was 0.14, and the death rate was 0.08. These means apply to the population as a whole. Imagine the real population: we do not have 0.14 of each animal giving birth. Rather, a proportion of the population reproduces and it is impossible to say if a given animal will survive or will have offspring in any year. To model this aspect of a population, we could ask if each member of the population survives and, independently, if each member has offspring in any time step. A time step of a year seems appropriate because reproduction in the species is seasonal and the environment, especially rainfall, is highly seasonal (Conway and Goodman, 1989).

We account for two things by implementing the model in this way. We treat the population as composed of an integer number of individuals and we sample the survival and reproduction of members of the population, using the observed population size and the population average birth and death rates. The result is that our predictions will no longer be exact. A run of bad luck sampling the survival of individuals could lead to the extinction of any population, no matter how large the population size or how large the potential growth rate. These components of uncertainty are termed demographic stochasticity (see Chapter 1).

This approach is most effectively implemented on a computer. Most programming languages incorporate a random number generator. Usually, they generate a random number from a uniform distribution scaled between 0 and 1 (that is, each value between 0 and 1 has the same probability of being sampled). The algorithm to calculate population growth is:

Algorithm 2.1 *Exponential population growth with demographic variance.*

1. For each time step from 1 to t, do steps 2 to 7.
2. Let $N(t + 1)$ take the value of the current population size, $N(t)$.
3. For each animal from 1 to $N(t + 1)$, do steps 4 to 7.
4. Choose a uniform random number, U_1.
5. Choose a uniform random number, U_2.
6. If U_1 is less than d, then decrease $N(t + 1)$ by 1.
7. If U_2 is less than b, then increase $N(t + 1)$ by 1.

Notice that the formulation in the Algorithm 2.1 is similar to, but not quite the same as, Equation (2.5), even if we ignore the element of demographic stochasticity it simulates. In the algorithm, we specify the average size of next year's population according to the equation

$$N_{t+1} = N_t + bN_t - dN_t \qquad (2.6)$$

Thus, Equation (2.5) says that $N_{t+1}/N_t = \exp(b - d) = 1.062$ whereas Equation (2.6) says $N_{t+1}/N_t = 1 + b - d = 1.060$. Neither of these formulations is wrong. Rather, what is meant by b and d is slightly different. Because the population reproduces seasonally, in a more or less discrete period rather than continuously as, say, humans do, the definition of b and d encapsulated in Equation (2.6) is appropriate.

Notice also that Algorithm 2.1 simulates each individual separately. All the algorithms in this chapter use this approach to simulate demographic stochasticity because it makes the process easy to understand. By using Algorithm 2.1 we stipulate that a female can have no more than one offspring per year, and if $b = 1$ there won't be any uncertainty associated with births. These assumptions are probably allright for white rhinoceros, but not for many other species. In the chapters following Chapter 2, a more efficient and more general method is used that involves sampling the binomial and Poisson distributions.

In formulating the model above, we assumed that births and deaths are independent events. Two components of Algorithm 2.1 reflect this assumption. First, in steps 6 and 7, if an animal dies we allow it to have offspring in that same year. Second, we choose different random numbers to represent the survival and reproductive success of each individual in steps 4 and 5. We could, if we wanted, preclude reproduction if an animal dies by returning to step 2 if U_1 is less than d in step 6. The version of the model in Algorithm 2.1 was implemented and the results of the first two trials are shown in Figure 2.2.

It is clear that demographic stochasticity can have an important effect

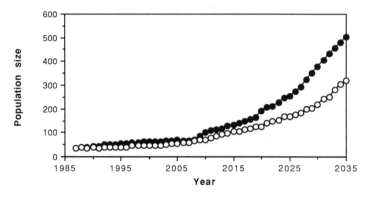

Figure 2.2   Two trials of a model for exponential growth of a white rhinoceros population incorporating demographic variance ($N_0 = 35$, $b = 0.14$, $d = 0.08$). The population size is the number of females.

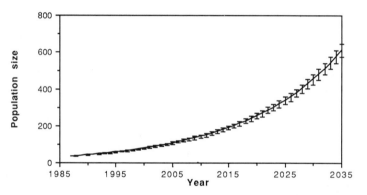

Figure 2.3   Mean and upper and lower 95% confidence limits for exponential growth of the white rhinoceros population, based on 100 replications of Algorithm 2.1. Population size is the number of females.

on estimated population size. This kind of variability is present in every population. The deterministic expectation (Figure 2.1) will be unlike most of the possible trajectories. It is fundamentally unrepresentative because it does not diverge from the pattern of smooth exponential increase. The second trial (Figure 2.2, open circles) falls well below the predictions of the deterministic version. We have done no more than take the trouble to treat our population as a finite set of integer numbers. If the trials are repeated many times, it is possible to calculate the mean and confidence limits of the population size at each time step. We simulated growth for the population 50 times using Algorithm 2.1 and performed these calculations for the above model (Figure 2.3).

The mean prediction of the stochastic model is similar to that for the deterministic model (Figure 2.1). Notice that the confidence interval increases in width as time goes on (Figure 2.3). Our predictions become less and less certain, the further into the future we make predictions. This characteristic is a general result common to all stochastic models and it makes good intuitive sense.

The 95% confidence limits in Figure 2.3 are quite close to the mean, even after 50 years. On the whole, demographic stochasticity has often been considered unimportant in models of all but the very smallest populations because it has such a small effect. Demographic stochasticity is likely to be most important for rare species because the variance in population size it causes is related directly to the number of individuals in the population. As populations become larger, the effects get smaller, a phenomenon we shall return to below. We can see the qualitative effect of population size by considering the survivorship probability, 0.08, in isolation. If two females remain, the chance of extinction due to variability in sampling is $0.08^2 = 0.0064$. When there are 60 animals, the chance is 0.08^{60}, a very small number. Nevertheless, there remains some chance of important deviations from the deterministic model (Figure 2.2) and some chance, albeit small, of extinction through nothing more than sampling error. We shall return to the relationship between variance and population size in Chapter 3.

2.5 INTRODUCING A POPULATION CEILING

Conway and Goodman (1989) note that the population on Ndumu Reserve is below the carrying capacity of the environment. They expect the population to continue to increase, and they plan to remove individuals whenever the population exceeds about 60 adults. Given a constant proportion of about 20% subadults and juveniles in the population, this population ceiling translates to about 40 females.

The animals taken are to be used in establishing or supplementing herds in other areas. Another reason for the removals is that the area of suitable white rhino habitat on the Reserve, although small, supports a large number of other grazers and increased rhino numbers could be detrimental to the chances of persistence of other species.

We can model this aspect of the management of the population in a straightforward manner by using Equation (2.6). We include a population ceiling for females above which there is no reproduction, and we truncate the population to 40 whenever it exceeds that number. The model becomes:

for $N_t \leqslant 40$, $N_{t+1} = N_t + bN_t - dN_t$

$$(2.7)$$

for $N_t > 40$, $N_{t+1} = N_t + bN_t - dN_t - H$

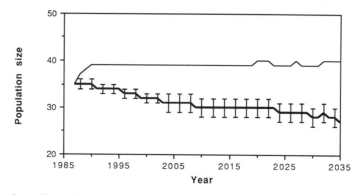

Figure 2.4   Size of the female rhinoceros population under two different scenarios. The thin line represents a management plan involving removal of animals only when the female population exceeds 40. The heavy line represents removing a number of animals that is proportional to 6% of the current population size, as well as animals whenever the female population size exceeds 40. The latter scenario may be interpreted as representing moderate to high levels of poaching. Both trajectories are based on 100 replications. The upper result is without confidence limits because they were too small to be drawn with accuracy. The error bars on the lower curve (removal of 6% of the population) are the 95% confidence limits. When just 2% of the population was removed, instead of 6%, the results were almost identical to the case of a simple population ceiling (the thin line).

where *H* is the harvest, equal to the number of females in excess of 40. This model may be implemented on computer using the same structure as Algorithm 2.1. The logic will be,

Algorithm 2.2   *Population growth to a ceiling with demographic variance.*

1. For each time from 1 to *t*, do steps 2 to 8.
2. Let $N(t + 1)$ take the value of the current population size $N(t)$.
3. For each animal from 1 to $N(t + 1)$ do steps 4 to 7.
4. Choose a uniform random number, U_1.
5. Choose a uniform random number, U_2.
6. If U_1 is less than *d*, then decrease $N(t + 1)$ by 1.
7. If U_2 is less than *b*, then increase $N(t + 1)$ by 1.
8. If $N(t + 1)$ is greater than 40, then let $N(t + 1)$ be 40.

This form of the model is equivalent to a management plan in which we only remove animals when the female population exceeds 40. The mean results of 100 replications of this model (Figure 2.4) show that the population is unlikely to deviate much from the imposed ceiling of 40 female rhinoceros. In fact, the 95% confidence limits are less than 1, even to the end of the simulation.

If managing this population involved nothing more than removing the excess animals each year, we would not anticipate too many problems from demographic stochasticity. However, even a modicum of intuition makes us feel uneasy about a population of 40 female rhinoceros surviving for the next 50 years. Too many things can happen.

Conway and Goodman (1989) mention the possibility of poaching, an activity responsible for the decline of rhinoceros populations elsewhere in Africa. Poaching may result in a constant number of animals lost from the population each year. Alternatively, if denser populations are easier to poach, it may be better expressed as a constant proportion of the population lost each year. Taking this latter alternative, we may rewrite the model in Equation (2.7). We let p be the per capita rate of loss due to poaching. That is, it is the average chance that an animal has of being taken by a poacher in any one year. Equation (2.7) becomes,

$$\text{for } N_t \leq 40, \qquad N_{t+1} = N_t + bN_t - dN_t - pN_t,$$
$$\text{for } N_t > 40, \qquad N_{t+1} = N_t - dN_t - pN_t. \tag{2.8}$$

We make essentially the same assumptions about poaching as we do about births and deaths. The rate of poaching is independent of the other two events, it is independent of the age, size or sex of the individual, and it is constant in time. Again, using the same basic algorithm, we can write the logic for Equation (2.8) as:

Algorithm 2.3 *Population growth to a ceiling, with demographic variance and removals proportional to current population size.*

1. For each time from 1 to t, do steps 2 to 9.
2. Let $N(t + 1)$ take the value of the current population size $N(t)$.
3. For each animal from 1 to $N(t + 1)$, do steps 4 to 9.
4. Choose a uniform random number, U_1.
5. Choose a uniform random number, U_2.
6. Choose a uniform random number, U_3.
7. If U_1 is less than d, then decrease $N(t + 1)$ by 1.
8. If U_2 is less than b and $N(t)$ is less than 40, then increase $N(t + 1)$ by 1.
9. If U_3 is less than p and U_1 is greater than d, then decrease $N(t + 1)$ by 1.

We have assumed that the chances of dying, having offspring, and being taken by poachers are independent events for each animal. We have no data on the real levels of poaching on the Reserve. Assume that poaching occurs at a modest rate, say, one animal per year from a population of 50. In this case, the value of p will be around 0.02. Note from Equation (2.8) that the effect of poaching at this rate is identical to increasing the death rate from 0.08 to 0.10. The reason for the extra condition in step 9 in

Algorithm 2.3, 'if U_1 is greater than d', should be clear enough. If U_1 is less than the death rate, d, then that particular animal has died. If an animal both dies and is taken by poachers in the same year, it amounts to the same thing as far as the population is concerned. The condition in step 9 avoids double counting the animal.

The average population size and confidence limits based on 50 simulations show that population size remains close to the ceiling of 40 females under the modest poaching rate of 2% per year. If the poaching rate is increased to 6% (equivalent to taking three animals out of 50 per year), the population size becomes very much more variable and the mean population size falls to less than 30 females after 50 years (Figure 2.4).

2.6   REMOVING CONSTANT NUMBERS

We know that rhinoceros are difficult to count, so it is unlikely that a management programme will be able to remove a constant proportion of the population. To do so, we would need to know the population size at the time the animals are removed. Instead, it is more likely that a specified number of animals would be removed each year. Conway and Goodman (1989) suggest the population at Ndumu could sustain the removal of two individuals (or one female) per year. Poaching may work essentially the same way if the number of animals taken is independent of the current population size.

Using the algorithms already developed as a template, it is easy to write a further algorithm for removing a set number of animals. For simplicity, we shall dispense with the population ceiling. Letting X be the number of animals taken, Equation (2.6) is modified to become

$$N_{t+1} = N_t + bN_t - dN_t - X. \tag{2.9}$$

The algorithm becomes:

Algorithm 2.4   *Exponential population growth with demographic variance and removal of a constant number of animals per year.*

1. For each time from 1 to t, do steps 2 to 7.
2. Let $N(t + 1)$ take the value of the current population size $N(t)$, less the number removed, X.
3. For each animal from 1 to $N(t + 1)$, do steps 4 to 7.
4. Choose a uniform random number, U_1.
5. Choose a uniform random number, U_2.
6. If U_1 is less than d, then decrease $N(t + 1)$ by 1.
7. If U_2 is less than b, then increase $N(t + 1)$ by 1.

We simulated this population model over 50 years, and repeated each trajectory 100 times. These simulations were performed twice each,

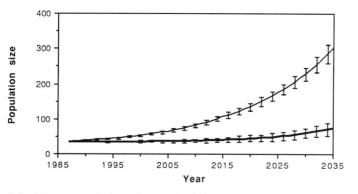

Figure 2.5 Mean population size and 95% confidence limits based on 100 replications for the rhinoceros population, assuming one female (thin line) and two females (heavy line) are removed each year.

letting X equal 1 and then 2 to represent the removal of either 1 or 2 females each year from the population.

The results suggest, at first glance, that the population could sustain the removal of two females per year and remain viable (Figure 2.5). The mean population size is higher, after 50 years, than the mean population size if we impose a strict ceiling (Figure 2.4). If it saves considerable time and money not to survey the population each year immediately prior to removing animals, we may be content to remove as many as two male and female pairs every year.

2.7   ENVIRONMENTAL VARIATION

So far, we have looked at nothing except the variation in population size that results from sampling a finite, integer number of animals from one year to the next. Populations are affected by other things, notably by the variation in the environment around them. Conway and Goodman (1989) observe, for example, that extended drought has probably depressed natality in the population of rhinoceros at Ndumu. Such variation, we assume, is fundamentally unpredictable.

Changes in environmental conditions will dictate fluctuations in the growth of a population. That is, environmental variation will affect the population through effects on survivorship and fecundity. When conditions are poor, food and water are scarcer, there is more competition with other grazers for the available resources, and fewer rhinos survive the year. Thus, environmental variation may be seen as something that varies the population mean birth and death rates.

Perhaps the easiest way to model this in a system without density dependence is to incorporate environmental variation through a degree of random variation in the parameters for growth rate, b and d. As an equation, this may be written as

$$b_t = b + s \cdot y_t \tag{2.10}$$

and

$$d_t = d - s \cdot y_t \tag{2.11}$$

where b and d are the mean birth and death rates respectively of the population, y_t is a random variable with a mean of zero and a variance of 1, and s represents the standard deviation of the fluctuations. In the long run, the average of the y values is 0, so the average b_t is b and the average d_t is d.

We have assumed that variations in the birth rate and the death rate act in equal and opposite directions. If the random number is greater than zero such that the birth rate increases, then the death rate decreases. This means that if it is a good year for the survival of adults then it is also a good year for the survival of juveniles from birth to the time of the first census. We could have done it in other ways. It is unlikely that the two parameters for birth and death are as perfectly correlated as they are in Equations (2.10) and (2.11). We could choose different random numbers for each parameter, resulting in a correlation of zero between them. We could correlate the two random numbers so that if it is a good year for juveniles, it is likely (but not necessarily) a good year for adults. The correlation may even be negative if, for some reason, a good year for the survival of adults results in poorer survival of animals from birth up to first census. This will be true if the adults compete for the same resources as the juveniles, as adults are better competitors, and resources are limited.

The values corresponding to y_t are often assumed to be 'white noise': this term means that the random distribution from which y_t is drawn is the same at all times and there is no correlation between values selected for successive times. If species abundances are the result of several factors that act independently, the effect of these factors on population size will be additive. The Central Limit Theorem suggests that environmental variation acting in this way may best be represented by the normal distribution. If these independent factors act in a multiplicative fashion (i.e. their logarithms are additive), then the approximate distribution will be lognormal (for an introduction to the normal and lognormal distributions, and the Central Limit Theorem, see Sokal and Rohlf, 1981).

To represent these ideas, we must consider the structure of our model so far. The birth and death rates represent, for each animal in the population, the mean chance per year of having offspring or surviving. In our previous models, we have assumed them to be constant values. We are now saying these rates are not constant in time, and that they may vary depending on how favourable or unfavourable the seasons turn out to be.

We shall assume that the environmental factors important for the growth of the rhinoceros population act independently and that their effects on the growth rate of the population are additive. For example, if rainfall is 10% above average, we might expect survival to increase by 10%. If some other important factor is 10% better than normal, we shall add another 10% to the average chance of survival.

All we need to do, at each time step, is to choose a random number from the standardized normal distribution with a specified standard deviation. This number will have a mean of zero. An algorithm for selecting a random number with a given standard deviation from the normal distribution is given in the Appendix, Section A.2.3. If the value we choose is negative, it represents a bad year for the population. The mean birth rate will fall and the mean death rate will rise (Equations (2.10) and (2.11)). Conversely, if the number is positive, it represents a good year, and mean birth rate rises and mean death rate falls. The magnitude we specify for the standard deviation represents the magnitude of the effects of environmental variation on the birth and death rates.

If we use the model for exponential growth in discrete time (Equation (2.2)) as an example and substitute Equations (2.10) and (2.11), it becomes

$$N_{t+1} = N_t + (b + s \cdot y_t)N_t - (d - s \cdot y_t)N_t \qquad (2.12)$$

where the value y_t represents environmental conditions between times t and $t + 1$, and s represents the magnitude of their effect on the population.

The degree of correlation we employ must depend on empirical data. In the absence of any other information, we assume that the parameters have a correlation of 1 (that is, we use the same number to represent variation in both parameters). It would be better to assume there is no correlation between births and deaths if the factors that act on births are different from those that act on deaths. To solve Equation (2.12) and predict population size, accounting for both demographic and environmental variability, we need only introduce one more step to Algorithm 2.1, giving us

Algorithm 2.5 *Exponential population growth with demographic and environmental variance.*

1. For each time from 1 to t, do steps 2 to 8.
2. Choose a normal random number, Q_1, with mean 0 and specified standard deviation.
3. Let $N(t + 1)$ take the value of the current population size $N(t)$.
4. For each animal from 1 to $N(t + 1)$, do steps 4 to 8.

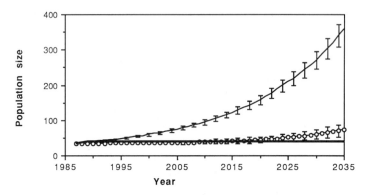

Figure 2.6 Predictions accounting for demographic and environmental variation for the size of the female rhinoceros population under three management scenarios: removal of animals when the population exceeds 40 (heavy line), removal of two females per year (circles) and removal of one female per year (thin line). The curves for the first two options are almost identical until about year 2015.

5. Choose a uniform random number, U_1.
6. Choose a uniform random number, U_2.
7. If U_1 is less than $(d - Q_1)$, then decrease $N(t + 1)$ by 1.
8. If U_2 is less than $(b + Q_1)$, then increase $N(t + 1)$ by 1.

To be realistic, b_t and d_t are truncated so they remain within the range [0..1]. The same applies to Equations (2.10), (2.11) and (2.12). Take two forms of the model developed above, Equations (2.7) and (2.9). These are, respectively, the equations for exponential growth to a ceiling and exponential growth with removals of a constant number of animals. The form of the equations, and the associated algorithms for the implementation of (2.7) and (2.9), may be modified in the same way as for the exponential model above (Algorithm 2.5). In Algorithms 2.2 and 2.4, just choose a normal random number (step 2 above), and modify the two steps for birth and death to include the variable Q_1.

We don't know anything about the levels of environmental variability on Ndumu Reserve, or how they affect the reproduction and survival of individuals in the rhinoceros population. Assume that the effects of the environment are moderate. We shall use a coefficient of variation of 10% in birth and death rates. The mean birth rate is 0.14. Random variation of this magnitude will mean that in 95 years out of 100, the birth rate will be between 0.11 and 0.17, due to the effects of the environment. A coefficient of variation of 10% in death rate means that, 95 years out of 100, the death rate will be between 0.05 and 0.11. Using these parameters, it is unlikely that the mean population death rate will exceed the mean population birth rate in any one year, due to the effects of the environment alone.

For the exponential model with environmental and demographic variance, the expected (mean) population size remains the same as the deterministic model. The curves for all three scenarios (Figure 2.6) look very similar to the predictions for the same models accounting for demographic variance alone (Figures 2.4 and 2.5). The addition of environmental variance had done little more than increase the 95% confidence limits slightly. None of the management alternatives seems likely to drive the population to extinction.

As managers of the rhinoceros population, we may be tempted to look at Figure 2.6 and conclude that removing constant numbers from the population is a reasonable option. The growth rate, on average, is positive and the confidence limits when removing both one and two females are a long way from zero. While the variability expected as a result of imposing a ceiling is much smaller, the mean population size is well below the other two options at the end of the period. However, the confidence intervals shown in Figure 2.6 may be misleading. The distribution of the population sizes is strongly skewed, particularly towards the end of the simulations. For example, in the case involving the removal of 2 females per year (Figure 2.6), the mean population size at the end of 50 years is about 70 females, with confidence limits of about 20. However, 39 of the 100 trajectories (and the mode) are zero at 50 years, and most have population sizes of less than 15 females. A few trajectories reach population sizes in excess of several hundred, reflecting the small chance that environmental conditions are consistently good for most of the 50 year period. Thus, the confidence intervals underestimate the true magnitude of the variation. We shall test these conclusions by adding risk assessment to our stochastic models.

2.8 RISK ASSESSMENT

The only thing that may cause us concern about removing constant numbers of animals is that this management practice results in a much more variable prediction of the population size. The prediction of the model for the option involving removals when the population exceeds 40 females, that is, the option for a population ceiling, results in almost no important variation. It is too small to be shown on Figure 2.6. We should be comfortable in knowing that the population size will be close to 60 adults if our model is even approximately correct.

When we calculate the mean and confidence limits from a set of numbers, we are looking at the central tendency of our data. The behaviour of outliers is largely ignored. Estimating risk involves calculating the chances of extreme events. Extreme events in any population may result in population crashes or explosions. In the case of the rhinoceros on Ndumu Reserve, we want to know the chances that the population will become extinct at some time in the next 50 years.

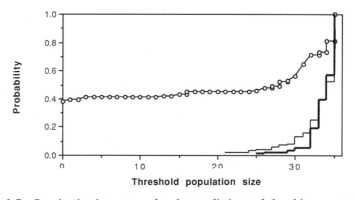

Figure 2.7 Quasiextinction curves for the predictions of the rhinoceros population size: ceiling of 40 females (heavy line), removing one female per year (thin
line), and removing two females per year (circles). The simulations are represented in Figure 2.6 and are based on 100 replications. To construct the risk
curves, the following procedure was used. From each replication of a simulation,
the smallest size attained by the population between 1986 and 2035 was recorded.
For example, the range of these values lay between zero and 35 for the simulation
involving the removal of two females. Then, for each population size over the
range of values (e.g. from 1 to 35), the proportion of the 100 values less than the
given size was calculated. This gave a cumulative probability distribution representing the chances that the population will fall below the given (threshold) size
at least once in the next 50 years. This process was repeated for each of the
simulations in Figure 2.6, giving the three risk curves above.

The simulations performed to generate the three trajectories in Figure
2.6 involved estimating the population size at each time step over 50
years, after sampling survival and reproduction to account for environmental and demographic uncertainty. From the set of 100 replications of
the simulation, it is possible to record not only the mean and confidence
limits for each time step, but also the smallest value calculated during
each replication. Each of the 100 trajectories that went into making
the three mean curves in Figure 2.6 were probably different (see, for
example, Figure 2.1). Some would have fallen below the mean trajectory
and some above, depending on the vagaries of our random number
generator.

To learn about the likelihood of extreme events, we need to look at the
smallest population sizes found from each replication of the simulation.
We may find that the predicted population size falls below, say, 20
animals, in a total of 40 out of 100 replications. We could then say there
is a 40% chance that the population will fall below a population size of 20
at least once within the next 50 years.

If we record the smallest population size from each simulation, we can
construct a graph of probabilities versus population size. A graph of this
kind is known as a quasiextinction curve (Ginzburg *et al.*, 1982) because it

represents both the chances of extinction and the chances of falling below small population sizes. The smallest population sizes were recorded for each replication of the three simulations in Figure 2.6, giving the three curves in Figure 2.7.

Perhaps the most surprising thing about these analyses involves the removal of two females per year. Even though the mean population size for this management option is well above zero, and above the ceiling of 40 animals at the end of 50 years (Figure 2.6), there is about a 40% chance that the population will become extinct (Figure 2.7). These risks are a much more effective guide to the likely future of the population than are the mean trajectories in Figure 2.6. They reflect the very strongly skewed distribution of the underlying population. Thus, even though the average population size for the situation involving the removal of 2 females is 70 at the end of 50 years, the chance that the population is less than 70 females is about 80%. We occasionally use confidence limits when displaying expected population sizes in the following chapters even though they do not give an accurate representation of the true amount of variation, especially in cases where the risk of extinction is high. The safest option is to implement a ceiling and remove animals only when the population exceeds it. The chances that the population will fall below 30 females within the next 50 years is small. Even removing just 1 female per year results in an increased risk that the female population will fall below 30 animals. These risks are incurred by the population, not because of systematic pressure (in the sense of Shaffer, 1987), but because the variability in the average rate of population growth is greater when constant numbers are removed.

There is a good reason why the removal of constant numbers poses a greater threat to the population than does the imposition of a ceiling. When we remove animals it may be thought of as mortality, at least from the point of view of the rhino population. Removing animals in excess of 40 is the same as increasing the chances of mortality when the population becomes large. When the size falls below the ceiling, mortality rates effectively decrease because we no longer take animals. Thus, managing the population in this way has a stabilizing effect on the population size. We shall tend to push the population of females towards 40 individuals from above and below.

If we decide, on the other hand, to remove two females per year, our management activities will destabilize the population, making it more prone to population crashes and extinction. This is because the effective mortality rate (the sum of per capita death rates due to natural causes and removals) will increase when there are fewer rhinos. For example, if there are 100 females in the population, the effective mortality due to removals will be 2%. If there are 10 females, it will be 20%. This management scheme makes mortality inversely proportional to the density of the population and will tend to push it towards extinction.

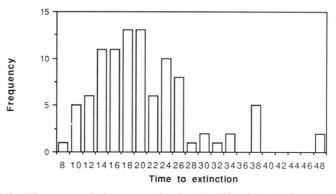

Figure 2.8 Histogram of time to extinction in 100 trials of the simulation of exponential growth of the rhinoceros population. The model includes demographic and environmental variance, and three females per year are removed from the population. A total of 97 replications out of 100 became extinct within 50 years and the remainder were extinct within 100 years. Only those that became extinct during the first 50 years are shown. The labels on the *x*-axis are lower class limits. Thus, the class labelled '8' represents all cases where the rhinoceros population became extinct during either the 8th or 9th year of the simulation. The label '20' contains the count of all replicates that went extinct during the 20th or 21st years.

2.8.1   Interpreting estimates of risk

If we find congruence between a model's predictions and observations of a population in the real world, it provides only weak inductive evidence that the assumptions of the model are correct. When models are tested against the real world, often we talk of model predictions. Predictions are of two kinds. Most often they refer to nothing more than the statistical fit of a model to a set of observations. The prediction of a regression model is the line that provides the least squares best fit to the data. In general, the more parameters one has in a model, the more variation of the data will be explained by the model. The second type of prediction is a forecast, wherein a model is used to predict events or values measured in the future. The parameters of a model are estimated from independent observations, and the forecast may be tested. Here, we have been forecasting the size of the rhinoceros population, but our interests are qualitative. The model is by necessity very simple because our data are limited.

A better model for these animals could include more factors to bring it nearer to reality. For example, it is unlikely that it would go unnoticed by the wildlife managers if the population size were to fall to 30, half its current size. If it did so, they would not continue to remove animals. In our models for constant removal, we continued to take rhinoceros, even when the population was close to extinction.

Conway and Goodman (1989) planned to remove two animals per

year. The effect of their management recommendations is likely to be a mixture of the model involving removals of a constant number per year, and the model for a population ceiling. We could build this into our computer model by, for example, allowing removals only when the population exceeds 40 animals. There are doubtless many other factors we could include if our knowledge was a little better.

Nevertheless, we have learned a few useful things. First, removing two females per year results in a more variable population size than removing females in excess of 40, even if the population's average net production is more than two females per year. This, in turn, increases the risks that the population will decline through chance events in survival and reproduction, and through a succession of poor years. Second, while wildlife managers are likely to be sensitive to the risks of declining population size, poachers are not. These results tell us that if poachers remove a constant number of two females or more per year, the population is endangered. If they remove as many as three animals per year, the population is almost certain to become extinct within the next 50 years (Figure 2.8). Poaching seems to pose a real threat because, even if moderate, its intensity may be independent of the existing population size. Poaching will destabilize the population. Careful management around the desired population ceiling may stabilize the population size and thereby reduce the chances of a severe reduction in numbers over the next 50 years.

The problems of the managers are complicated by measurement errors. Since they may not know the exact number of animals, sometimes animals may be removed when $N < 40$ and not removed when $N > 40$. This will increase the variation in population size and partially (even completely, if estimates of N have large errors) offset the stabilizing property of this management strategy. Its effectiveness depends on how precisely N is estimated.

When we simulate the scenario where three females per year are removed, the distribution of extinction times is skewed to the right and, as a result, most of the trials resulted in the population going extinct before the mean extinction time. About 60% of the replicates became extinct before the mean extinction time of 21.5 years. This example shows why it is important to know the complete distribution of extinction times, not just the average time (see Chapter 1).

The model described above is used as an exploratory tool. We have been interested principally in the qualitative results of different assumptions. We do not really expect the population to be 379 females, plus or minus 35, in 50 years' time if we remove exactly one per year. Too many things may happen in the intervening period to make the prediction realistic. The Reserve may expand or contract in area; we may develop a technique that makes counting rhinoceros easy; a catastrophe such as extreme drought or disease may eliminate the population in a single year.

These things are not accounted for in the model we developed. They may change our approach to modelling, or they may allow the implementation of different management practices. We do, however, expect that if we manage the population by implementing a ceiling, the population has a better chance of persistence than if we remove a constant number of animals. It appears to be important to remove animals only if we know the current population size.

2.9 SUMMARY

The processes of birth and death in a population determine changes in population size. The so called birth-and-death models can be used to predict population sizes at various times in the future. A model is constructed for white rhinoceros in Ndumu Reserve, providing an example of how to build a stochastic model.

The exponential model of population growth after 50 years predicts a population size greater than could be supported by the environment. Demographic variability adds an important component of uncertainty to the model's predictions. A modification to the model allows the evaluation of different management practices to limit the size of the population. Population variability will be minimized if animals are removed only when their abundance exceeds a specified threshold, but the effectiveness of this strategy depends on the errors associated with estimating population size. If constant numbers are removed from the population every year, regardless of its current size, the effect is destabilizing because the proportion removed increases as the population size decreases.

Environmental variation may be introduced to demographic models through the parameters for population mean birth and death rates. The effects of different impacts on the rhinoceros population can be estimated through the risks of population delcine. Removing constant numbers from the population greatly increases the chances of extinction of the population.

3 Useful methods when data are scarce

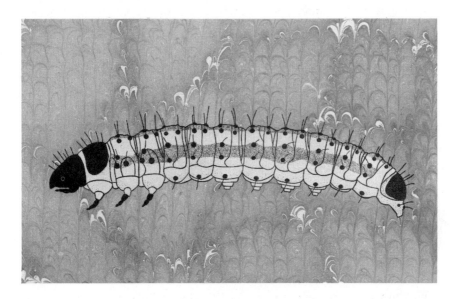

It is possible to build a workable model even when there are gaps in your knowledge of a population. The exact values for things such as reproductive rates, population densities, or other parameters may be unknown. The important thing is to build a model that reflects the detail of knowledge about a population. For some large mammals (like African lions and many primates), and a few economically important insects and plants (such as the giant kelp, *Macrocystis pyrifera*), there are years of painstaking observations on their population sizes, behaviour and dynamics. For these species it is possible to construct large and complex models to make firm predictions. They are not necessarily better predictions than those made from a two-parameter equation.

When detailed data are missing, as they are for most species, it would be fatuous to attempt to build a complex model. The information at our disposal for the majority of species is likely to be patchy and estimates of parameters may be little more than educated guesses. In Chapter 2 we built a model for the population dynamics of white rhinoceros, knowing little more than the population size, population ceiling and the mean birth and death rates. The simplest models, such as the one developed for the rhinoceros population, treat the population as a single number; that is, they ignore demographic and spatial structures within a population. Because of their utility, this class of models, which are called single-dimensional or unstructured models, have been well developed in the literature. They have been used widely to model biological populations and they do not rely on a great deal of information.

In this chapter we describe unstructured models and present several examples to show some of their advantages and some of their dangers. We begin with a more complete treatment of the exponential growth equation developed in Chapter 2. Much of the chapter is devoted to some of the most important and widely used of the equations for unstructured models, including the logistic and several other general equations for density-dependent population growth. We introduce the notion of non-linear components in equations. We outline sensitivity analysis and present a general algorithm for population growth. To illustrate some of the features and limitations of these models, we develop models for populations of shrews (*Crocidura russula*) from suburban gardens in Switzerland, magpie geese (*Anseranas semipalmata*) from the tropical floodplains of northern Australia, and larch budmoths (*Zeiraphera diniana*) from the mountains of central Europe using some of the best known of the unstructured population equations.

3.1 THE EXPONENTIAL MODEL FOR POPULATION GROWTH

The exponential model appears superficially of little value because few populations inhabit environments in which they may grow indefinitely. Eventually, the resources that individuals require for reproduction and survival will be depleted. However, when populations are small relative to the abundance of resources and amount of empty habitat, we could expect them to grow at a rate close to the maximum. In fact, most models for which we provide the details below predict exponential growth over part of the range of population sizes.

In Chapter 2, we used the exponential model for population growth. It has been applied to natural populations of plants and animals, both because it is simple and because it fits a number of data sets for population growth quite well. The exponential model will be an adequate tool to describe a real population and predict future population sizes if the density range is that in which exponential growth may be expected. In most cases, this will be when the population density is low compared to the carrying capacity of the environment (e.g. in cases where available habitat has increased or harvesting and predation pressure have been reduced).

Predictions from the exponential model
Eberhardt (1987) collected population estimates from several
species of large mammals. The estimates were taken over periods
ranging from 5 to 19 years. They included marine animals (grey
whales, sea lions and four species of seals), white-tailed deer
from three different areas, elk from two areas, Serengeti buffalo,
feral horses, bison, muskox and longhorn cattle. The exponential
growth model provided an excellent fit in many cases, although
it was necessary to introduce a functional representation of density dependence in some instances. Eberhardt noted there may be
striking differences in the way density dependence operates in
superficially similar species such as longhorn cattle and bison.

3.1.1 Environmental variation

In a real population, changes in environmental conditions will dictate
fluctuations in the growth of a population. In most cases, these changes in
the environment will be unpredictable. In the example in Chapter 2,
we incorporated environmental variation through a degree of random
variation in the parameters for the birth rate and death rate (Equations
(2.10) and (2.11)). Alternatively, we could make the growth parameter,
r, a stochastic parameter. As an equation, this may be written

$$r_t = r + s \cdot y_t \tag{3.1}$$

where r is the mean growth rate of the population, y_t is a random variable
with a mean of 0 and a variance of 1, and s represents the standard
deviation of fluctuations. Instead of treating the birth rate b and the death
rate d separately as we did in Chapter 2 (Equation (2.12)), we summarize
the effects of the environment on them through the parameter r. For the
exponential model in continuous time with environmental variance, the
expected (mean) population size remains the same as the deterministic
model (Maynard-Smith, 1974, pp. 13–14); that is,

$$N_t = N_0 \, e^{rt} \tag{3.2}$$

where N_t is the population size at time t, N_0 is the initial population size,
and r is the instantaneous growth rate of the population. This result
assumes that variations in growth rate around r in Equation (3.2) are
normally distributed. For more information on the effect of variation on
the expected size of an exponentially increasing population, see Levins
(1969) and Lewontin and Cohen (1969). If environmental variance is
included in the model through a degree of variability in r, then the
variance in the expected population size at a given time in the future

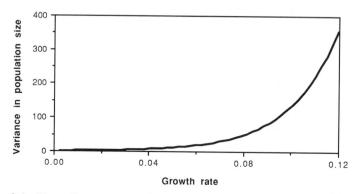

Figure 3.1 The effect of mean growth rate (*r*) on the expected variance of population size in the exponential growth model (Equation (3.3)). Mean growth rate was varied and all other parameters were held constant. The values for variance in the mean population size are those expected at the end of 25 years.

can be calculated. It is given by the equation (Maynard-Smith, 1974, pp. 13–14),

$$\text{var}(N) = N_0 \, e^{2rt}(e^{s^2t} - 1). \tag{3.3}$$

We can use this equation to demonstrate the effect that growth rate and variability in growth rate will have on the certainty with which we can predict future population sizes. The larger the variance, the less certain will be our predictions. We used the parameters for the white rhinoceros population from Chapter 2: the initial population size was 35 animals, and we assumed a standard deviation in growth rate of 0.1. The mean population growth rate was varied and the effect on variance in population size calculated.

Variance in population size increases exponentially with increasing *r* (Figure 3.1). As *r* increases, the population responds more rapidly to a given change in the environment. Population size becomes more and more difficult to predict for a given level of environmental variation. By inspection of Equation (3.3), we can see that the variance in population size increases with time; that is, our confidence in the estimate of the population size will fall, the father into the future we make predictions. The same phenomenon was evident in all of the stochastic simulations for the rhinoceros population undertaken in Chapter 2. The confidence limits for the predicted population sizes broaden with time. The same is true of variance in growth rate. The larger the variance in the environment, the larger will be the expected variance in population size.

Often, we shall want to know the chances that a population will fall below some specified size. This is the risk of quasiextinction (Ginzburg *et al.*, 1982; see Chapter 1). In this case, we are interested in estimating the

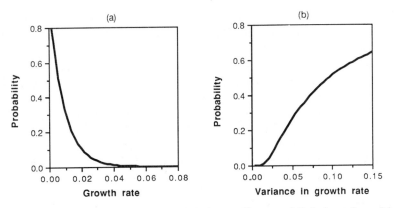

Figure 3.2 Probability that the populations will ever fall below the critical threshold of $N_c = 20$ individuals as a function of (a) growth rate, and (b) variance in growth rate. The curves were calculated from Equation (3.4). For (a) we assumed $N_0 = 35$, $s^2 = 0.01$, and r varied between 0.002 and 0.08. For (b) we assumed that $N_0 = 35$, $r = 0.06$ and s^2 varied between 0.0036 and 0.15.

risk that a population will ever fall below a threshold size, N_c, given an initial population size, N_0, and a level of variance, s^2, in the growth rate, r, that is a result of variability in the environment. The probability of crossing a threshold for the exponential model is

$$P = \left(\frac{N_c}{N_0}\right)^{2r/s^2}. \tag{3.4}$$

We can observe a number of things about the qualitative effect of population growth rate and the variance in population growth on the chances of persistence of a population from this equation. In the example in Figure 3.2 we have assumed, arbitrarily, that the critical lower size for the population is 20 animals and that the population grows exponentially from an initial size of 35 animals. Equation (3.4) gives us the chances that the population will ever fall below 20 animals (the probability of quasiextinction of the population) for a range of mean growth rates (Figure 3.2(a)) and variances in growth rates (Figure 3.2(b)).

The probability of quasiextinction declines asymptotically to 0 as the growth rate increases (Figure 3.2(a)). The faster the growth rate, the faster the population is carried away from the critical boundary of 20 animals, and the smaller are the chances of crossing it. It only reaches 0 when the growth rate is infinitely large, so there will always be some chance of quasiextinction, no matter how large the potential rate of increase of the population.

The probability of quasiextinction increases asymptotically to 1 as the variance in growth increases (Figure 3.2(b)). The larger the variance, the larger the fluctuations in population size around the mean rate of increase, and the greater are the chances that the population will cross the boundary of 20 animals. The probability reaches 1 only when the variance is infinitely large, so there will always be some chance that a population will persist, no matter how variable the environment.

These qualitative results translate from quasiextinction probabilities to extinction probabilities. The only difference is that as the threshold boundary is reduced towards a population size of zero, the probabilities of crossing the boundary are reduced correspondingly.

Equation (3.4) (Figure 3.2) underestimates extinction risks because it ignores demographic, genetic and spatial uncertainty. For small numbers such as $N = 20$ or $N = 35$, demographic and genetic uncertainty may have important effects.

Ginzburg et al. (1982) suggest using the relative change in the probability of quasiextinction as a measure of impact on a population. To do this, we need to measure the probability of extinction or quasiextinction of a population, both with and without the foreseen impact. This index is written

$$I = \frac{(P_{imp} - P)}{P} \qquad (3.5)$$

where P is the probability of extinction in the absence of the impact. The risk faced by a population in the absence of an impact is termed the background risk of quasiextinction. This index is scaled from 0, for no impact, to infinity. It may be used to compare the impacts of management alternatives, where I is an index of the amount of risk added to the population as a result of the proposed actions.

There is no established agreement on using relative increase in risk (Equation (3.5)) as a measure of impact. It could be expressed in an absolute scale (i.e. $I = P_{imp} - P$) or as a ratio ($I = P_{imp}/P$). For example, an impact that increases risk of extinction from 1 to 4% is unlikely to be judged as severe as one that increases it from 25 to 100%. It may be that an absolute scale is inappropriate as well. An impact leading to an increase from 1 to 4% may be considered less severe than one that leads to an increase from 97 to 100%. Relative scales of impact will probably be useful if the background risks (the starting point) are the same. In this case, different measures of impact will imply the same conclusions, so it will not matter which is used.

You may wonder why the model in Chapter 2 is necessary when all these equations are available to calculate population sizes and extinction probabilities. The reason is that a computer program is much more flexible than these equations. It allows us to include almost any aspect of the biology or dynamics of the population, such as adding a ceiling of 40 females, just by adding two or three lines of code. While it is possible to find an analytical solution to the problem of a population ceiling (i.e. there are equations to predict population sizes and extinction probabilities for a population that grows exponentially to a ceiling), it is by no means easy to derive it. The program allows us the freedom to include any other variation in the model we may care to add, and we know we shall always be able to calculate population sizes and extinction risks.

3.1.2 Demographic variation

If we consider only demographic variability, it is possible to develop an analytical stochastic model for a population. These 'birth-and-death' models are based on the assumption that each individual in a population has a certain probability of death and a certain, independent, probability of reproducing during an interval of time.

Bartlett (1960), Pielou (1977) and Goodman (1987b) provide derivations for several birth and death models. It turns out that the expected population size at time t is identical to the population size predicted in the deterministic growth equation (Equation (3.2)) above (Pielou, 1977, p. 16). The model also provides an estimate of the variance in the expected population size. This equation is given by several biomathematicians including Kendall (1948, p. 5), Poole (1974, p. 51) and Pielou (1977, p. 17) as

$$\text{var}(N) = N_0 \frac{b + d}{b - d} e^{(b-d)} (e^{(b-d)t} - 1). \tag{3.6}$$

The variance in the expected population size (at some specified time, t) represents the differences in population size at time t, resulting from different replicates begun with population size N_0. It is an exact solution to the estimate of variance obtained from running the simulations in Chapter 2, for the model for exponential growth with demographic variance.

As for the equations for environmental variance above, the equations for demographic variance are useful to establish a few generalizations concerning demographics, population dynamics and extinction risks. The variance from demographic uncertainty (Equation (3.6)) depends on the difference between the birth and death rates. The larger the difference, the larger will be the variance. This means that we can predict the population sizes of species that increase very rapidly with less confidence than we can predict the population sizes of species that increase more slowly.

Interestingly, the variance depends on the absolute magnitudes of the birth and death rates. If the rates of birth and death in a population are large, the estimate will be less precise than those made in a different population with smaller rates, even if the intrinsic rate of increase, the difference between these parameters, is the same. It should be noted that when the birth and death rates increase, but the difference remains the

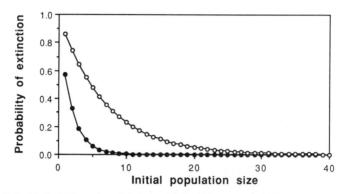

Figure 3.3 Probability of extinction within 100 years as a function of the initial population size, as a result of demographic stochasticity. For the open circles, birth rate = 0.44 and death rate = 0.38. For the closed circles, birth rate = 0.14 and death rate = 0.08 (the latter numbers are the vital rates for the white rhinoceros population on Ndumu Reserve; Chapter 2). The two curves were generated from Equation (3.7). The mean population growth rate in both cases is 0.06.

same, the turnover rate of individuals in a population increases, so there are more chance events per unit time.

Of interest in cases where populations are very small or declining is the question of the probability of extinction within some specified time. Again, within the framework of a 'birth-and-death' model and considering only demographic stochasticity (Pielou, 1977, p. 19),

$$p(t) = \left(\frac{de^{(b-d)t} - d}{be^{(b-d)t} - d}\right)^{N_0}. \tag{3.7}$$

This equation calculates the probability, p, of the population becoming extinct by some specified time, t, given the initial population size, N_0, and the population mean birth and death rates.

When considering demographic variance, the chance of extinction in an exponentially growing population depends on the same factors that affect the variance in the expected population size (Equation (3.7)). The chance of extinction is increased by a small intrinsic rate of increase, large absolute values for birth and death rates, and a small initial population size.

In the two curves in Figure 3.3, the population mean growth rate (i.e. the difference between the birth and death rates) is the same. Nevertheless, the chances of extinction are quite different. The population with higher mean rates is much more likely to become extinct. When considering demographic stochasticity in a small population, the ratio of birth rate to death rate is much more important than the difference between birth and death rates. In Figure 3.3, the ratio of b/d for the population that is more likely to become extinct is 1.15, compared to 1.75 for the other population. The probability of extinction due to demographic stochasticity is very small for the vast majority of cases when population size exceeds about 20 or 30 individuals for simple models of birth and death.

3.1.3   General results

Equations (3.1) to (3.7) have two important advantages over computer simulations. First, they are faster and easier to compute. Second, the solutions they provide are exact. When we estimate confidence limits from a set of simulations, we are using a sample from the very large number of possible trajectories for the population. We should use a large number of replications, especially if, for example, the confidence limits around the mean are broad.

The probability of extinction increases as a function of the magnitude of the variance in the growth rate (Figure 3.2). Further, as the variance in growth rate increases, probability of extinction or quasiextinction becomes less sensitive to a reduction in growth rate. These results make it clear that it is insufficient to measure the impact of management procedures in terms of the reduction in population growth rate without considering their effect on the level of variability in growth rates.

Equations (3.4) and (3.7) imply one result that is perhaps counter-intuitive. There remains a finite risk of extinction, even in a population of very large size growing exponentially in an unvarying, unlimited environment. If the variance in growth rate is more than twice the mean growth rate, the probability of extinction tends to unity as time tends to infinity; that is, if fluctuations are large enough, populations whose expected size is increasing will nevertheless ultimately become extinct. Lewontin and Cohen (1969) and Maynard-Smith (1974) provide more details on this aspect of population growth.

3.2   DENSITY DEPENDENCE, THE LOGISTIC EQUATION AND MAGPIE GEESE

All populations grow within environments where eventually they will become limited by a shortage of resources. Given a long enough time span, the assumption of constant mean birth and death rates made in an exponential model could break down. The observation that most populations appear to be present in more or less constant numbers from year to year has led to the development of two views of population regulation.

The first is based on population dynamic equations incorporating density dependence. The idea is that there exists a *deterministic* relationship between the abundance of a species and the abundance of available resources or the condition of their environment. Put another way, the population growth rate, r, depends on N through its effect on resource availability. Descriptions of equations for this relationship may be found in many good ecology textbooks (see, for example, May, 1976b, Pielou, 1977, and Krebs, 1985).

The second view of population regulation emphasizes the *stochastic* relationship between a population and its environment: populations do not grow for very long before the environment changes and they begin to decline. Population size is, in general, dominated by perturbations of the environment (see Andrewartha and Birch, 1954, p. 19; Strong, 1986). Population growth depends on changes in environmental factors, including changes in resources, but not through their changing availability as a result of changes in N. For example, a wide variety of animals and plants may be affected by a single detrimental environmental event. Unusual spring weather and unseasonal cold in Colorado, USA, in 1969 caused extensive damage to herbaceous perennials, insects and small mammals and caused the local extinction of at least one butterfly population (Ehrlich *et al.*, 1972).

The notion that population growth is limited by the amount of available habitat and resources, and the relationship between these limits and variation in the environment, is important for understanding and predicting the fate of populations. In Chapter 5, we present a framework wherein the spatial dynamics of several populations is used to replace the equilibrium dynamics of single-population models.

The magpie goose (*Anseranas semipalmata*) is a wildfowl that lives in tropical northern Australia. Populations are abundant on subcoastal wetlands where they inhabit ephemeral rivers and floodplains. The life cycle of the goose is dominated by the driving environmental variable, rainfall. During the dry season (May to October), the geese congregate around permanent waterholes and in the wet season (November to April) they disperse across the floodplains to nest.

The density of goose populations appears to have an upper limit, perhaps related to the number of available nesting sites. In any case, in each of the populations sampled, the growth rate of the population slows when density is high. Bayliss (1989) chose the logistic equation, the most thoroughly studied model of density-dependent population dynamics, to represent the dynamics of the magpie goose populations and we use his results as an example of its application to natural populations.

Density dependence is defined as the phenomenon by which the values of vital rates such as survivorship and fecundity depend on the density of the population. For example, if a bird population inhabits an environment with a limited number of nesting sites and most of these sites are taken, relatively few offspring will find themselves a vacant nest the following year. When adults are few, a larger proportion of offspring will find nests and the rate of population increase will be greater.

The simplest assumption to make is that the mean growth rate of the population decreases linearly with increasing population density. That is,

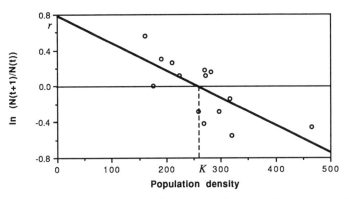

Figure 3.4 Density-dependence function for magpie geese. Carrying capacity, K, is 260 animals, maximum growth rate, r, is 0.78 (after Bayliss, 1989, his Table 1; this model ignores one outlier). The slope of the line, $-c$, is given by $-r/K$.

if population size doubles, the remaining resources are depleted at a rate proportional to the population increase. Either fecundity or survival, or both, decrease at a rate proportional to the decline in unused resources. Taking the exponential growth equation ($\mathrm{d}N/\mathrm{d}t = rN$), we may divide both sides by N and add a constant, c, to represent the rate at which growth rate decreases with increasing population size. The average growth rate of the population is given by

$$\frac{1}{N}\frac{\mathrm{d}N}{\mathrm{d}t} = r - cN. \tag{3.8}$$

Bayliss (1989) was interested in the relationship between population growth rates of magpie geese, population density and variation in rainfall. Using data from aerial photographs, he calculated the rates of increase of populations between consecutive years from the ratio of $\log_e(N_{t+1}/N_t)$, and regressed these values against the current population size (N_t). Although this approach is widely used, and we use it below, the regression is not strictly valid. The x and y axes are not statistically independent because the value on the x-axis (N_t) appears as the denominator in the corresponding value on the y-axis. This lack of independence means that a negative slope is more likely than a positive one, even from a series of random numbers, because an increase in N_t decreases N_{t+1}/N_t.

The y-axis of Figure 3.4 represents the realized growth rate of the population. The point at which the line intercepts the y-axis is equal to r, the maximum average growth rate of the population in the absence of density-dependent effects. We label the point at which the line intercepts the x-axis, K. At this population density, the number of individuals exactly consumes available resources and the average growth rate is zero. The number of births exactly equals the number of deaths in the goose populations when the population density reaches 260 animals per square kilometre. At that point, the left-hand side of Equation (3.8) equals zero and we can see that

$$c = \frac{r}{K}. \tag{3.9}$$

If population density exceeds K, deaths exceed births until such time as the population returns to K. Substituting Equation (3.9) into Equation (3.8) gives the standard equation describing these dynamics

$$\frac{dN}{dt} = rN\left(1 - \frac{N}{K}\right). \tag{3.10}$$

Here, K represents the carrying capacity of the environment, N is the population size and r is the maximum instantaneous rate of population increase. Equation (3.10) is the Verhulst–Pearl logistic equation. Pielou (1977, pp. 21–2) describes several independent lines of argument that lead to it. Like the equation for exponential population growth, the logistic equation makes several assumptions about the underlying dynamics of a population. They include:

1. The population has a stable age distribution (that is, the proportion of individuals in each age class does not vary from year to year; see Chapter 4).
2. In the continuous time form, the response by a population to an increase in density is instantaneous.
3. The intrinsic rate of increase is reduced by a constant amount for every individual added to those already present.
4. Crowding affects all individuals and life stages of a population equally.
5. The environment is constant.
6. Demographic and genetic effects are unimportant.

The right-hand side of Equation (3.10) may be evaluated to give the rate of change in population size as a function of the intrinsic rate of natural increase, r, the current population size, N, and the carrying capacity, K. By inspection, we can see that if the population size is less than the carrying capacity, the term inside the parentheses will be positive and the population will increase. If the population size is greater than the carrying capacity, the term will be negative and the population will decrease. Thus, the equation is said to have a 'globally stable equilibrium' at $N^* = K = 260$ (the asterisk (*) is used to indicate the equilibrium value of a variable). The populations will always tend to return to a density of 260 animals per square kilometre.

Equation (3.10) can be solved by integration, giving an equation for N_t as a function of the carrying capacity and the rate of increase. Given an initial population size, N_0, the population at time t is,

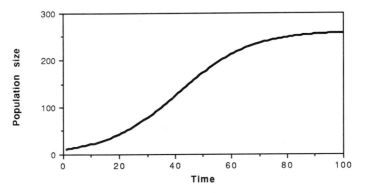

Figure 3.5 Logistic growth for magpie goose populations. Population size as a function of time was estimated using Equation (3.11) with $K = 260$, $r = 0.78$ and $N_0 = 10$.

$$N_t = \frac{K}{1 + [(K - N_0)/N_0]e^{-rt}} \, . \tag{3.11}$$

This model will preclude the possibility of extinction as long as r is non-negative and K is positive. Using this equation, we can predict the population size for Bayliss' goose populations at any time t in the future, knowing that $r = 0.78$, $K = 260$ and given some initial population size, N_0 (Figure 3.5). If the initial population is below the carrying capacity, the change in population size follows a sigmoid pattern to K.

3.2.1 Discrete time

Equations (3.10) and (3.11) may be appropriate if the population reproduces and dies continuously, as do human populations. However, when population growth is markedly discontinuous, as it is in the magpie goose populations, a difference equation is more appropriate. A review of numerous population dynamic equations in discrete time is provided by May and Oster (1976).

Discrete models offer a few advantages over their continuous analogues. Many habitats and the dynamics of many species of plants and animals are strictly seasonal. Moreover, by using time units of a season or a year, we give the system we model sufficient time to homogenize (Lomnicki, 1988, p. 101); that is, there is a time lag between the deaths of individuals and their replacement by reproduction that is ignored in continuous models. If the time step in a model is short, we must model more closely the behaviour of individuals. To model populations in continuous time we shall require more data and better understanding of the way in which the population works. Lomnicki (1988, p. 102) suggests that detailed models in continuous time are not particularly useful when data are scarce.

One of the most widely used discrete time versions of the logistic equation is,

$$N_{t+1} = N_t\left[1 + r\left(1 - \frac{N_t}{K}\right)\right]$$ (3.12)

It is important to understand that while Equation (3.12) looks a lot like its continuous time analogue, Equation (3.10), it is really very different. The most important difference is that the continuous form of the logistic always leads to smooth population changes. A population starting below the carrying capacity will follow a smooth sigmoid trajectory of the sort shown in Figure 3.5. The discrete form of the logistic can result in very complicated population abundance curves. The dynamics observed will depend on a number of things, including the size of the time step used, and the magnitude of the growth rate, r.

We can examine the differences between discrete and continuous logistic models by looking at recruitment curves. Recruitment curves represent the population size expected at time $t + 1$, given the number present at time t. The underlying difference between continuous and discrete forms of the logistic is that the discrete equation uses a time step of a given length. Between one time and the next, the dynamics of the population are governed by the conditions in the population at the start of that time step, and by nothing else.

Consider a magpie goose population in which the population density is very high, say, 520 birds km^{-2}. We decide to census the population at the end of the dry season when birds are concentrated around permanent waterholes and are easy to count. We use the discrete logistic model because births are restricted to a single pulse, the period during the wet season when floodplains are inundated. We use a time step of one year because the dynamics of the population are governed by rainfall, and rainfall is highly seasonal. Equation (3.12) predicts that the population at the next census would plummet to 114 birds (Figure 3.6). From there, the population will recover gradually to the equilibrium density of 260 birds. If we had instead used Equation (3.10), we would predict that the population size will fall to about 340 birds at the next census and will decline gradually to the equilibrium density over the next few years.

Equation (3.12) has some interesting characteristics. Population abundances will vary in the way they approach the carrying capacity, K, depending on the magnitude of the population growth rate, r. If r is less than 1, population size increases in a manner entirely analogous to the continuous version of the model. As r increases, the population's approach to K changes qualitatively. The population size will successively undershoot and overshoot the carrying capacity, resulting in damped, or undamped, oscillations. When r is large enough, predictions for population size become chaotic, which means that even though the equations are

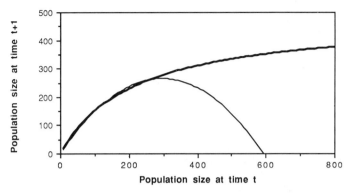

Figure 3.6 Recruitment curves for the magpie goose populations. The parameters of the model are the same as those used in Figure 3.5. The curves
represent the number of individuals expected in the population at time $t + 1$,
given the number present at time t, and are calculated for continuous time using
Equation (3.11) (heavy line) and for discrete time using Equation (3.12) (thin
line).

entirely deterministic, very small changes in the initial population size
will lead to wildly different predictions. Discussion and descriptions
of these dynamics may be found in Maynard-Smith (1968), May (1976b,
pp. 11–14) and Krebs (1985, pp. 210–11).

This kind of instability is characteristic of many discrete models. The
logistic equation predicts that a population will become extinct in the next
generation if the population size ever exceeds $K(1 + r)/r$. If the magpie
goose population exceeds 593 birds km^{-2} (the point of interception of the
recruitment curve with the x-axis, Figure 3.6), Equation (3.12) predicts
extinction by the time of the next census. The continuous analogue will
not be unstable, to matter how large the growth rate. Large populations
will always return smoothly to the carrying capacity. The magpie goose
example shows that a logistic equation in discrete time is very different to
its continuous time analogue. If you want to approximate a continuous
time model with discrete equations, so that you can simulate population
growth on a computer, you should use the Ricker function with $\beta = r/K$
(see Section 3.3).

3.2.2 The logistic model with environmental variance

Bayliss (1989) recorded the density and rate of population increase of
magpie geese together with rainfall. The results of statistical analyses
showed that rainfall in the previous wet season is important in determining population densities. Thus, the populations are at least partially
dependent on year-to-year variation in rainfall. Even if we identify the

correct model to use for these populations, and the estimates of mean population birth and death rates and the carrying capacity are accurate, predictions of future population sizes must be uncertain because the amount of rainfall is not predictable.

The logistic equation has two parameters, r and K, and environmental variation may be introduced realistically by varying either of them. To account for the effect of variable rainfall on the survival and reproduction of the geese, we assume the deviations around the population mean growth rate at any given density that result from variation in rainfall are approximately normally distributed. The model for population growth we use is developed by substituting Equation (3.1) into Equation (3.12), making r a normal random variable:

$$N_{t+1} = N_t + N_t(r + s \cdot y_t) - \frac{rN_t^2}{K}. \tag{3.13}$$

The magnitude of the variation in r is given by the standard deviation, s, and it represents the magnitude of environmental fluctuations. The parameter y is a standard normal deviate. Only the first term with r varies, otherwise once the population reaches carrying capacity, there would be no variability in population size. The second r in Equation (3.13) does not vary because it is part of the term representing density dependence. Arranged this way, the equation implies that the amount of environmental variation a population experiences is independent of how close the population is to the carrying capacity. To understand this more fully, consider using the alternative formulation,

$$N_{t+1} = N_t \left[1 + (r + s \cdot y_t) \left(1 - \frac{N_t}{K} \right) \right]. \tag{3.14}$$

When the population reaches the carrying capacity, the term $(1 - N_t/K)$ goes to zero, and any variation associated with r vanishes. Furthermore, the closer we are to the carrying capacity, the more the variation in r is diminished. There is no biological intuition that would suggest that the loss of variation near K is likely for the goose populations, so Equation (3.14) is inappropriate. Using Equation (3.13), we can account for environmental variation without assuming that the magnitude of variation depends on population size. The populations of geese on the floodplains of northern Australia are large. Even before we perform any analyses, we know there is very little chance the populations will fall below many thousands of individuals within the next 50 years. Figure 3.3 tells us that the added risk due to demographic stochasticity will be diminishingly small for populations as large as this, and we can safely ignore it.

If we begin with a population at the carrying capacity of 260 birds km^{-2}, and we use the discrete deterministic form of the logistic equation

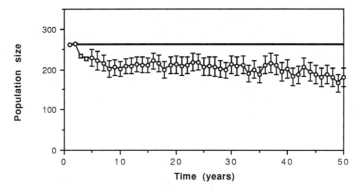

Figure 3.7 The logistic model for growth of a magpie goose population in northern Australia. Initial population density was 260 birds km^{-2}, and maximum growth rate, r, was 0.78 year^{-1} (after Bayliss, 1989). Population size is expressed as birds km^{-2}. The unbroken line represents the predictions of the deterministic logistic equation (Equation (3.12)). The open circles are the mean population size from 100 replications of a simulation for Equation (3.13) assuming a standard deviation in growth rate of 0.3, the same magnitude of variation used by Bayliss. The error bars are the 95% confidence limits of the stochastic prediction.

(Equation (3.12)), we predict that the population will remain at 260 birds km^{-2} indefinitely (Figure 3.7). We get the same result if we use Equation (3.13) and set s equal to zero. Setting s equal to 0.3 in Equation (3.13) gives qualitatively different predictions for the average population sizes over the next 50 years (Figure 3.7).

The most noticeable thing about Figure 3.7 is that the mean population size of the stochastic model is lower than the predicted population size from the deterministic version of the same model. The explanation for this phenomenon may be found by looking at Equation (3.13). Growth rate is controlled by the difference between the terms N and N^2/K, which means the rate of approach to K is faster from above than from below. There will be a greater lag recovering from low densities than there will be when the population returns to the carrying capacity from high densities (Whittaker and Goodman, 1979). This characteristic alone is sufficient to reduce the mean population size of the stochastic model below that of the deterministic analogue.

Another reason for the difference is that the population size of zero acts as an absorbing barrier. Once a population is extinct, it cannot recover. We would therefore expect the mean population size to decline gradually as more and more of the replicates become extinct. There were no extinctions among the replications for Figure 3.7.

This study exemplifies one of the most important differences between analogous stochastic and deterministic models. The logistic equation is non-linear because the population growth rate changes as population density changes. If a stochastic model embodies non-linear components, the mean (i.e. the expected) population size will be different to that predicted by the deterministic version. The stronger the non-linearity, the greater will be the difference (Chesson, 1978). This does not mean that the deterministic version of the model is wrong. Some populations may be driven largely by systematic, deterministic pressures. Environmental variation may have little effect on survival or reproduction. However, it is important to be aware that deterministic models will wrongly estimate population size if the system is non-linear and variability plays a role in the population's dynamics.

3.2.3 Harvesting

Magpie geese are good to eat and, although it is a protected species, some of them are taken by local people. Bayliss (1989) suggested implementing a system whereby the number taken in any one year is a constant proportion of the total available. Assume that each individual in the population has an equal chance of being taken. If the population remains well below the carrying capacity, we can then assume that the population will grow exponentially. One of the most common ways to calculate sustainable yield is to use the exponential model to predict the harvesting rate a population can sustain. For example, Flipse and Veling (1984) applied it to predict acceptable harvesting rates for the hooded seal (*Cystophora cristata*) in the north Atlantic (cf. Burgman and Neet, 1989). It was of interest to Bayliss to calculate the maximum sustained yield of the populations, the maximum number of geese that could be taken from the populations on a regular basis. This aspect of wildlife management can be investigated by introducing a harvesting term to the logistic equation. Using Equation (3.13) it results in

$$N_{t+1} = N_t + N_t(r + s \cdot y_t) - \frac{rN_t^2}{K} - HN_t \qquad (3.15)$$

where H is the proportion of the population alive at time t that is harvested between time t and time $t + 1$. By varying H between 0 and 1 and observing the density to which the population falls, we may calculate the number of geese we can expect to take under the entire range of harvesting intensities.

The striking feature of Figure 3.8 is that the stochastic model predicts substantially lower maximum sustained yields. The deterministic model predicts a maximum yield of 51 geese km^{-2} when the harvest rate is 0.40 per year. The stochastic model predicts a maximum yield of 38 geese when the harvest rate is about 0.35 per year.

Bayliss (1989) investigated the effects of harvesting a constant percentage of the geese, and likewise found that the potential sustainable harvest was much lower under the stochastic version of the logistic model than it

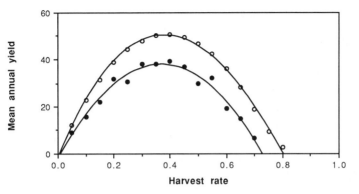

Figure 3.8 Mean annual yield of magpie geese (after Bayliss, 1989). Mean annual yield represents the average number of geese taken per km^2 during the 50th year of a simulation, allowing sufficient time for the population to stabilize close to new equilibrium values from the initial density of 260 geese km^{-2}. The open circles show predictions for the deterministic logistic equation. The closed circles show predictions for the stochastic model with standard deviation = 0.3. The trend lines are the third-order polynomials of best fit.

was under the deterministic version. The results were based on limited data but Bayliss commented that the predictions of the stochastic model rely on the strength of the relationship between growth rate and rainfall in the preceding year. Probably the most important effect of rainfall on the population growth of magpie geese is the inundation of nesting sites during the wet season (Whitehead and Tschirner, 1990). If the effects of floods turn out to be unimportant for population growth, the correct prediction may be more like the deterministic case. Other harvesting designs may be more appropriate for other species and Goodman (1982) provides formulas to calculate the removal rate of a fixed fraction of individuals from subsets of age classes (see Chapter 4), so as to achieve zero growth rate.

This example serves to illustrate some of the dangers inherent in models of natural populations. When data are scarce, it is unwise to make exact quantitative predictions based on these models. We should trust only the qualitative results. For example, we may overestimate the number of geese we can take from the population if we ignore environmental stochasticity. If it is part of our job to establish a limit to the number taken per year, we should approach the problem conservatively. Begin by taking, say, 5% and slowly increase the proportion taken each year thereafter. If the yield differs significantly from the mean prediction of the stochastic model, re-evaluate the model, looking for biological reasons why the population diverges from the assumptions of the logistic. Once they are found, modify the model and try again.

3.2.4   General results

An exact analytical expression has been found for the risk of quasiextinction for the logistic model in which r varies randomly. The probability that a population with an initial size N_0, will fall below a critical level, N_c, within a time, t, is given by Ginzburg et al. (1982). Unfortunately, it uses a logistic model equivalent to Equation (3.14), in which the magnitude of the variation in population growth depends on how close the population is to the carrying capacity. When $N = K$, there is no variability in the population due to variation in the environment. This rather artificial restriction results in very small extinction probabilities, even for small populations that experience random variation in r.

Most analytical treatments of the logistic model use the assumption that the growth rate, r, is varied by sampling it from the normal distribution. Ginzburg et al. (1982) assume that the growth rate is sampled from the normal distribution with independent samples taken at each time step. This introduces another qualification to our conceptual model of the dynamics of the population. We must assume that there is no temporal correlation between random environmental changes.

Despite the restrictive formulation of Ginzburg et al.'s model, there are a few interesting generalizations that can be made from it. Using numerical examples for the equation above, Ginzburg et al. (1982) found the probability of passing below some critical level increases with time for each fixed growth rate and variance. The higher the variance in growth rate, the less sensitive is the probability of quasiextinction to a reduction in growth rate. The longer the time, the more influential is the effect of a given reduction in growth rate. The probability of quasiextinction is very sensitive to level of the threshold chosen. The probability decreases very rapidly with decreases in N_c/K.

As in the density-independent model (the exponential), variance in population size increases with time. Levins (1969, 1970) has shown that environmental fluctuations reduce mean population density below that reached in the analogous deterministic model. The magnitude of the reduction is a function of the magnitude of the variance in the random parameter. If fluctuations are large enough, extinction will result.

The logistic equation has some problems, especially when r is a random variable. An important part of the formulation is the definition of K, which is not an independent parameter that reflects the environment but merely a descriptor of the equilibrium population size. It depends on the other parameter, r (Kuno, 1987). When $N > K$, the population declines to this equilibrium. However, the rate of decline is greater for higher r; a higher rate of increase (r) does not lead to a slower decline, but to a faster one (Pollard, 1981). This is because the same parameter, r, describes two things that are biologically unrelated: rate of growth at low population density, and rate of return to equilibrium after disturbance.

The growth rate, r, is used as a random variable representing the effects of environmental variation. If r becomes less than zero as a result of such environmental variation, then dN/dt is positively correlated with N (Fulda, 1981). If, in the case when $r > 0$, N is temporarily above the equilibrium (due to environmental variation or immigration), explosive growth approaching infinity results, which is certainly meaningless (Hallam and Clark, 1981).

Rather than introducing environmental variation through the growth rate, r, Roughgarden (1975) varied K and found that the variation of the population size is a function of the responsiveness of the population to changes in K (i.e. the magnitude of r) and the variability of K itself. Whittaker and Goodman (1979) sampled K from values representing three different degrees of environmental severity. They found the mean time to extinction to be inversely related to the population growth rate. The relationship between growth rate and mean population size in their model is complex, depending on the severity of the environment and the presence of refugees from extreme conditions within a species' range. Strebel (1985) likewise used variations in the carrying capacity, K. He showed that extinction probabilities are increased when there is a resonance between the species' generation time and the average interval between the fluctuations in available resources.

3.3   OTHER FORMS OF DENSITY DEPENDENCE

Any, and perhaps all, of the assumptions of the logistic model may be violated by many natural populations. In the theory of density-dependent regulation of populations, the notion of dependence implies a reciprocal causal relationship between population density and the factors that determine it. That is, negative feedback exists wherein population density affects resource availability which, in turn, affects population density. Where the precise mechanisms of density dependence are poorly understood, a general model may be useful because it facilitates comparison of species or populations in terms of model parameters. Differences may be attributable to populations rather than to differences in model biases (Bellows, 1981).

Different forms of negative feedback have been classified into two qualitative types, a dichotomy which dates back to Nicholson (1954). They go by the titles of contest and scramble competition. **Contest competition** is compensatory. Resources are partitioned so that an individual obtains a parcel necessary for survival, or not. Individuals are either fully successful, or they are not successful. The result is that a certain number of individuals will survive, irrespective of the initial number in the population, and all of the limiting resources are committed to population growth. **Scramble competition** is overcompensatory. Resources are shared more or less evenly among all members of the

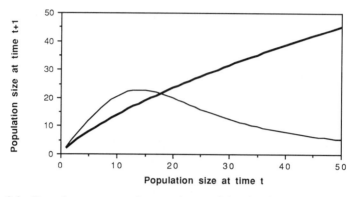

Figure 3.9 Recruitment curves for two forms of density dependence. The heavy line shows contest competition, the thin line shows scramble competition. Both curves were calculated using Equation (3.19), with $\lambda = \exp(r)$, $p = (\lambda - 1)/K$, $K = 25$ and $r = 0.88$. For contest competition, $q = 0.5$ and for scramble competition, $q = 3$. The value of λ equals the slope of the curve at the origin.

population. Some of the available resources do not contribute to population growth because they are won by individuals who do not survive. If some initial density is exceeded, fewer and fewer individuals in a population will survive because the share becomes smaller and smaller (see Lomnicki, 1988, pp. 20–45, 106–20).

Contest and scramble competition represent opposite ends of a continuum (Figure 3.9). Both forms may be represented in a single equation wherein the values of the parameters reflect how strongly the population responds to limited resources or habitat. A wide variety of alternative, unstructured density-dependence models were discussed by May and Oster (1976) and Hassell *et al.* (1976). Several general equations in discrete time, for which the logistic is a special case, have been proposed. One of these is given by Hassell *et al.* (1976) and May (1976a),

$$N_{t+1} = \frac{\lambda N_t}{(1 + aN_t)^b} \tag{3.16}$$

where $a = (\lambda - 1)/K$ and λ is the reproductive rate (e^r). This is the logistic equation (Equation (3.12)) when b equals 1.

A second equation that has been used widely in biological studies is the Ricker function (Ricker, 1975, p. 282),

$$N_{t+1} = N_t \, e^{(r - \beta N_t)} \tag{3.17}$$

The Ricker function always results in scramble competition, though its effects are less severe than the logistic equation. Note that it has only two parameters, r and β, rather than three as in Equation (3.16). The parameter r is dimensionless and β has dimensions of $1/N$. The value for λ, the growth rate of the population, gives the slope of the recruitment curve (Figure 3.9) at the origin while β is a measure of the strength of density dependence. The peak in the recruitment curve is reached when N_t equals $1/\beta$. As long as β is greater than zero, the form of the function will be a humped curve that asymptotically approaches zero at large N. The equilibrium population size, N^*, is r/β.

The response of the population is less severe as both parameters become smaller. Unlike the discrete logistic model (Equation (3.12), Figure 3.6), the Ricker function declines smoothly towards zero for high levels of reproductive effort but it does not reach zero. This feature may be important. If the environment is heterogeneous it may be unrealistic to wipe out all potential new recruits. A few may always survive because of a combination of good fortune and the haphazard distribution of resources.

A model that always results in contest competition is the Beverton–Holt function (Beverton and Holt, 1957; Ricker, 1975, p. 291). Like the Ricker function, it has two parameters,

$$N_{t+1} = \frac{1}{\rho + (k/N_t)} \ . \tag{3.18}$$

The parameters ρ and k take on non-negative values and ρ has units of reciprocal individuals. The term $1/\rho$ is equivalent to the carrying capacity of the environment. The parameter k is dimensionless and it determines the slope of the recruitment curve (Figure 3.9). At the origin, the slope is $1/k$. This is a useful model when there is a natural limitation to the recruitment of new individuals. If the survival of young is limited by the number of territories or the number of nesting sites, a more or less fixed number of young will be recruited, irrespective of the number of young produced. Any population dominated by these kinds of dynamics may be adequately described by the Beverton–Holt function.

A useful equation was developed by Maynard-Smith and Slatkin (1973),

$$N_{t+1} = \frac{rN_t}{1 + (pN_t)^q} \ . \tag{3.19}$$

It is superficially very similar to Equation (3.16). Here, p is inversely proportional to the amount of habitat or resource available (approximately $1/K$) and q controls the strength of the dependence of the population on available resources. It may be weakly linked if $q < 1$, or strongly linked if $q \gg 1$. When q is high, there are fewer survivors at high densities and scramble competition is implied. When $q = 1$, the number of survivors increases monotonically to a maximum of $1/p$. As the parameter p increases, the number of survivors at any given density decreases.

3.3.1 Choosing the right equation: dealing with the environment

We really need detailed autecological information to help us decide the correct functional form for a particular population. If the mating system or resource partitioning behaviour of individuals in the species suggests density-dependent limitations to population growth, we shall need a model that includes a density-dependence relation. When modelling a population that has become rare because of severe predation or a past catastrophe, you may conclude that the population is far from any other natural limitation on its size. If it is well below the carrying capacity of the environment, as are the rhinoceros in Chapter 2, it may be reasonable to model it without density dependence. If there is cannibalism or an increased likelihood of disease or onslaught of predators when population abundance is high, scramble competition is suggested. If breeding pairs use exclusive territories, and competition is manifested through competition for a limited number of territories, contest competition is suggested.

The problem then becomes, which density-dependence equation is best? In using these equations in computer simulations, one finds that population trajectories can vary quite sensitively, depending on how density dependence is modelled. In stochastic models, the chances of extinction and quasiextinction can be dramatically affected. Often, the data necessary to calibrate these models are expensive to collect or are not available (Ferson *et al.*, 1989). The differences in Equations (3.16)–(3.19) are subtle and we would need a great deal of information to decide which one is closer to a real representation of any population.

One approach when data are scarce is to choose the equation that has been found to provide the most reliable estimates of population size when used on other species with similar population dynamics. For example, Equation (3.17) is used regularly in fisheries biology research and Equation (3.19) has been used in fisheries research (Shepherd, 1982) and in models for terrestrial mammal populations (Dobson, 1988). This is not to say that if your species is a mammal, we recommend that you use Maynard-Smith and Slatkin's Equation (3.19). Very often, an equation is employed only because a program is available that features it, or the biologist is familiar with it. Very few comparative studies have been done on the relative merits and drawbacks of the different equations for population dynamics. Bellows (1981; cf. Bernstein, 1986) compared a number of them and found Equation (3.19) to be the general form with the greatest flexibility and best descriptive power, and the least prone to overestimate density-independent survival.

One important difference between scramble and contest competition is that contest competition always results in smooth, monotonic population dynamics whereas scramble competition can generate oscillatory and even chaotic dynamics. May (1989) distinguishes between 'density-dependent noise' generated by deterministic, non-linear relations, and 'density-independent noise' generated directly by environmental stochasticity. The processes of recruitment and survival are inherently stochastic because they depend on the vagaries of climate, resource availability and numerous other, unpredictable factors. The problems in deciding on the form taken by density-dependent regulation are exacerbated by environmental variability.

Variability is noise that obscures deterministic relationships. Non-linear population dynamics are often the most realistic, but they create special problems because they may result in deterministic fluctuations of population size. It is difficult in practice to distinguish between the variability that has deterministic origins and the variability that has environmental or demographic origins. A common definition of density dependence is that there is a negative correlation between the growth rate of a population and population size. The problem is that, for noisy data, it is very difficult to detect a slope of N_t/N_{t+1} that is different from unity.

A first step is to try to detect density dependence in the population census data. A number of methods for detecting it are available including k-value and key factor analysis (Figure 3.10). Strictly speaking, for key factor analysis we need detailed information on 'within generation' mortalities and their causes. However, it is difficult to know beforehand which parameters of the demography and the environment of a species are important in determining survival.

It has been found through empirical and simulation studies that it can be difficult to detect density dependence in natural populations because of the confounding effects of environmental and demographic variability (see Royama, 1977; Gaston and Lawton, 1987; Pollard et al., 1987; Ginzburg et al., 1990), and the spatial scale on which density dependence operates (see Hassell et al., 1987). For example, in studies based on computer simulations, Mountford (1988) found it possible to detect density dependence in the presence of environmental heterogeneity when population dynamics are governed by contest competition. However, Hassell (1986) found problems in detecting the form of density dependence when scramble competition was predominant (cf. May, 1989). Hassell et al. (1989) found density dependence easier to detect when censuses of populations were relatively long. Solow and Steele (1990) showed that for some models it is necessary to observe up to 30 generations before density dependence can be established confidently.

There are a number of other methods for detecting density dependence and Gaston and Lawton (1987) evaluated their effectiveness in cases where key-factor analysis was unsuitable. They conclude that none of the methods is a reliable test of density dependence in sequential censuses. Rather, it is important to have an understanding of the population dynamics and life history of a species, or carry out manipulative experiments on it.

Parameter estimates can be obtained by fitting the equation to observations using least squares techniques (e.g. Bellows, 1981; Ferson et al., 1989). One approach to choosing an equation is to fit a number of alternative equations to the census with non-linear regression: the equation that explains more of the variance in population size may be the best one. This approach treats deviations from the deterministic model as measurement error and the best equation minimizes this error. There is an example of this approach to the evaluation of models in Section 3.5.3 below. We have seen above that variability is a fundamental part of population dynamics and it may well be independent of the deterministic processes in a population. Statistical fit should be used with caution and should be underpinned with an understanding of the behaviour of individuals in the population. If a model does not make good biological sense, it should not be used to make predictions, no matter how well it fits a series of population sizes.

Regulation in fish populations

Fish populations have been the subject of a great deal of research, mainly because their economic value makes it important for us to predict future population sizes, particularly in the presence of fishing mortality. As a result, the data sets for many species of fish are considerably more extensive and more reliable than they are for many other natural populations. Shepherd and Cushing (1990) reviewed numerous studies on the role of density dependence in the regulation of fish populations. They found that the analyses of recruitment curves did not yield clear evidence for or against density-dependent regulation because of high levels of fluctuations in population numbers (see Ginzburg *et al.*, 1990; Fogarty *et al.*, 1991). Shepherd and Cushing conclude their study by suggesting that an important regulatory process in many fish populations may be a stochastic one: population sizes from year to year may become more variable when stocks are low.

Fogarty *et al.* (1991) evaluated the use of the exponential, Ricker and Beverton–Holt models in predicting exploited fish populations. They found that the probability of stock collapse increases with increasing environmental variation and increasing harvesting rate. They noted the difficulty in determining the form of stock-recruitment (density dependence) relationships for short time series, a point we address in detail below. Lastly, they found the practice of constant quota harvesting enhances the probability of population extinction, the same result we found in Chapter 2 with the model for constant removals from the rhinoceros population.

Many studies are confounded by considerable noise, such that little of the variation in population densities can be explained by models of the kind described above. Because of this unexplained variation, it is difficult to know whether populations are regulated by responses to declining

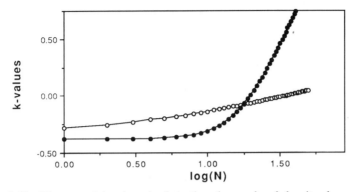

Figure 3.10 The use of *k*-values in detecting the mode of density dependence. The ordinate represents the *k*-value for the population and is calculated from $k = \log_{10}(N_t/N_{t+1})$. The abscissa is $\log_{10}(N_t)$ and *N* is the population size. Strictly speaking, the numerator in the expression for *k* is the population size immediately before the action of the density-dependence mechanism and the denominator is the population size immediately after its action. Begon *et al.* (1986) provide an excellent introduction to *k*-factor and key-factor analysis. Hassell (1986) and Hassell *et al.* (1987) give some examples. The slope of the right-hand side of the resulting plot, where it is more or less linear, suggests either scramble or contest competition. When the slope of the line is equal to 1, it denotes exact compensation. The closed circles show the expected relationship for scramble competition (slope $\gg$ 1); the open circles represent contest competition (slope $\leqslant$ 1). These curves were generated using Equation (3.19) with *b* set to 3 and 0.5, respectively. The calculation of *k* may be based on mortality, fecundity, growth or any other population parameter reflecting the response of a population to changes in density.

resources, or are more often limited by variations in climate and resources that are independent of population density. (See Krebs (1985) for an account of early debates; and Hassell (1986), Lomnicki (1987), Strong (1986) and Hassell and Sabelis (1987) for more recent debate.) The pervasiveness of stochasticity is borne out by the difficulty in detecting density dependence in real populations. In the absence of knowledge about the form of density dependence or the mechanisms determining it, it will be very difficult to determine the levels of regulation, habitat loss, or disturbance a population can sustain.

In short, there is no easy answer to the question of which equation to use or, once you have it, how to estimate the parameters correctly. There is no substitute for autecological information. We return to the problem of dealing with environmental variation in Section 3.5.5 below, where we introduce the problems associated with parameter estimation. The question is important to conservation biology because, as we shall see below, estimates of quasiextinction risks depend quite sensitively on how density dependence is modelled.

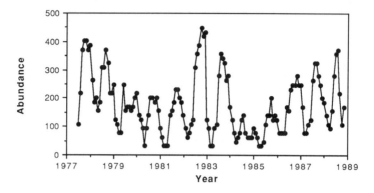

Figure 3.11   Size of the shrew population in western Switzerland (after Burgman *et al.*, 1992). Estimates of population size were made at approximately monthly intervals over 12 years. A calendar year is represented by the interval between vertical lines on the *x*-axis.

3.4   A MODEL FOR SUBURBAN SHREWS

Crocidura russula is a commensal shrew common in southern Europe, and the northern limit of its range extends as far as the North Sea. In western Switzerland, the species is found in fallow land, farmland and in residential gardens, living beneath hedges of thuya (*Thuya occidentalis*) and hornbean (*Carpinus betulus*). The animals have a life span of about one and a half years. In winter, they form communal nests with as many as six individuals, but during the reproductive season, pairs form and share common home ranges of about $70 \, m^2$, the females having consecutive litters with a mean of 4.4 young in each. In summer, the juveniles may either disperse or remain within the territory of the adults, generally reaching maturity at the end of the following winter. A population of *Crocidura russula* lives in suburban gardens at Preverenges in Switzerland, and there have been a number of studies on the reproduction and physiology of individuals from this population (Cantoni and Vogel, 1989 and references therein).

The area of potential habitat for this population is about $3000 \, m^2$. While it is not possible to know if it is completely isolated from other populations, no individuals have been observed entering the study area from elsewhere. For the purposes of this study, we assume the habitat within it is uniform, individuals may move freely between the gardens, and there is no immigration.

The population exhibits a strong seasonal cycle (Figure 3.11) and it is not immediately obvious which month to use to represent the annual abundance of the species. We treat the system on a one-year time step censused at the same time each year because the reproduction of the

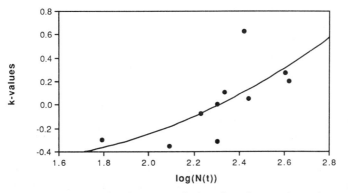

Figure 3.12   *k*-Values for the shrew population, based on estimated population sizes in November of consecutive years. The line is the quadratic equation that best fits these data, and the slope at the right-hand side is very close to 1. The *k*-values are explained in detail in the legend of Figure 3.10.

species is distinctly seasonal, restricted to the warmer spring and summer months. We use data from November (autumn) for the analyses here but the results are qualitatively similar if the data are taken from March (Burgman *et al.*, 1992).

3.4.1   Formulating the model

Subadult *C. russula* reach sexual maturity rapidly at the beginning of spring and population size is probably limited by the number of available territories (Besançon, 1984). Year-to-year variation in this population is density dependent and is mediated by contest competition, the key factor being the survival of young to adulthood (R. Arditi and D. Cantoni, unpublished data). Figure 3.12 shows *k*-values calculated from the ratios of estimated population sizes in consecutive years.

Two things tell us that density dependence is important in this population. First, Figure 3.12 seems to describe more than random noise. The slope of the line is close to 1 and there is no clear, sharp increase in the slope that would suggest scramble competition. The approximately linear relationship indicates compensatory dependence on available resources – that is, contest competition. Second, we know that this species forms home ranges during the reproductive season that are to some extent exclusive. Thus, we may expect partial partitioning of resources when shrews compete to establish a home range prior to breeding. This information suggests that we should expect a more or less fixed number of young to survive because the environment is divided up into parcels, home ranges, containing sufficient resources to support a breeding pair. This mechanism for resource partitioning is reminiscent of contest competition. While these dynamics may be adequately summarized by the Beverton–Holt equation, (3.18), if there is some degree of scramble competition in the survival of young to maturity, then the Ricker function (Equation (3.17)) may be a better descriptor. It is more flexible because the strength of the scramble competition may be weak, closely approximating contest competition at all but the highest population densities.

While it is true that shrews have home ranges, repeated trapping of the animals in the gardens over many years has shown that these ranges are by no means completely exclusive territories. Suppose that our first intuition is incorrect and that while the ranges have a role in the breeding behaviour of the species, competition for food is an unmitigated scramble. Overlap in home ranges could promote scramble competition if densities become very high. Such a mechanism will give rise to population growth rates adequately described by the discrete logistic equation. On the other hand, we could choose the discrete logistic equation because it is commonly available and well known, and therefore unlikely to meet much resistance from journal reviewers. For the sake of illustration, we will also model the population using it.

3.4.2   A general algorithm for density-dependent population growth

Most non-linear models, especially those with stochastic elements, are very difficult to solve analytically. While there are solutions for the logistic equation, there are none available for most of the other equations described above. We shall arrive at solutions for stochastic versions of the logistic equation, (3.12), and the Ricker equation, (3.17), using a discrete time numerical simulation (Algorithm 3.1). The algorithm to calculate the mean, confidence limits, chances of quasiextinction and mean time to extinction or quasiextinction for a single population is basically the same, no matter what model is used. It is very similar to those used to solve the models in Chapter 2 and it was used to predict population sizes for the magpie goose population in Section 3.2 above.

To implement the algorithm, you will need to have an equation that describes population size at time $t + 1$ as a function of population size at time t. We shall assume that the mean survivorship, S, is fixed independently of population size and that density dependence is mediated through fecundity. Thus, to calculate demographic stochasticity, you will need to supply a mean adult survivorship value of between 0 and 1. We use the binomial distribution to sample survivorships and the Poisson distribution to sample the number of young produced.

Algorithm 3.1 Density-dependent population growth incorporating environmental and demographic variability

1. For each time from 1 to t, do steps 2 to 6.
2. Calculate $N(t + 1)$ using the version of your model that accounts for

environmental variance, for example,

$N(t + 1) = N(t) + (r + Q_1)N(t) - r(N(t)^2/K)$ for the logistic.

Q_1 is a normal random number with mean 0 and specified standard deviation.

3. Let M equal the value $(N(t + 1) - S.N(t))$.
   If M is less than 0, set M equal to 0 and set S equal to the ratio $N(t + 1)/N(t)$.
4. Draw the number of young produced, Y, from a Poisson distribution with mean M.
5. Draw the number of survivors, A, from a binomial distribution with S as the probability and $N(t)$ as the sample size.
6. Let $N(t + 1)$ equal the sum of $A + Y$.
7. Record the smallest (or largest) value of N among the t time steps.
8. If the population crossed the critical threshold, record the time step in which it did so.
9. Repeat steps 1 to 8 a large number of times.
10. Calculate the mean value of N at each time step.
11. Calculate the confidence limits for N at each time step.
12. Output the histogram of times to crossing the specified threshold.
13. Calculate the chances of crossing the lower (or upper) bound from the array of smallest (or largest) values.

Algorithm 3.1 is flexible. If we want to omit demographic stochasticity because, for example, the population is never small or environmental variability is very large, just omit steps 3 to 6. If we want to account for it, these steps let us avoid simulating the birth and death of each individual in a population. This speeds up simulations of population dynamics considerably.

In steps 3 and 4 we assume that S is constant and that any variation in population size due to environmental variation acts on fecundity, M, unless population declines are relatively large. If it makes more sense for both fecundity and survivorship of a species to vary because of environmental variation, the reduction in step 3 can be made on both M and S. It will then be necessary to decide the amount of the variation that affects each of the parameters.

The values supplied for the mean and the variance of the random parameters, or the value of any of the parameters or variables in the model, may be functions of time. You could substitute the equation in step 2 with any appropriate formula. Any statistical distribution may be used to generate the random deviate for environmental variation. If the algorithm is used to simulate population growth in continuous time, choose a time step that is very small relative to the generation time of the species, and adjust time-dependent parameters such as growth rate accordingly. The information recorded allows calculation of the mean and confidence limits for the population, the mean time before crossing a threshold, and the chances of crossing the threshold within a specified time.

It is beyond the scope of this book to provide computer code to implement the algorithms described. The code for sampling the binomial and Poisson distributions, as well as many other distributions, may be found in books that provide recipes for numerical simulations. Good recipes may be found in Press *et al.* (1986). We provide some fundamental principles for numerical simulation in the Appendix, Sections A1 and A2. For discussion of the use of the Poisson and binomial distributions for demographic stochasticity, see Brillinger (1986) and Akçakaya (1990b).

3.4.3 Parameter estimation for the discrete logistic equation

Once the model has been formulated, initial conditions and estimates of parameters must be supplied. Normally, they include such things as the initial population size, growth rates, parameters for density-dependence functions, the population size representing the critical lower (or upper) threshold, and the magnitude of the variance of the random variable(s). Unfortunately, when census data alone are available, it is not possible to estimate birth and death rates independently. Because the data are unavailable and because the population was not small in November over the last 12 years (Figure 3.11, cf. Figure 3.3), we shall ignore demographic stochasticity.

The logistic model of density dependence implies a linear relationship between population size and growth rate (see Equation (3.8) and Figure 3.4). The realized growth rate for population size N at time t can be estimated from observed data for two successive years from the relationship,

$$r_N = \frac{1}{N} \frac{dN}{dt} \approx \ln \left(\frac{N_{t+1}}{N_t} \right). \tag{3.20}$$

The slope and intercept of the line for Equation (3.8) represent the rate of change in population size as a function of current population growth rate, and the maximum population growth rate in the absence of density-dependence effects, respectively. Thus, the parameters of the logistic equation can be estimated using linear regression. We fitted this regression relationship to the shrew population data, using the density of females alone to estimate the parameters of the logistic equation.

The regression of growth rate on population size (Figure 3.13) suggests that the shrews have a maximum potential growth rate of 0.917 in the absence of any density-dependent effects. The carrying capacity in the logistic equation is the point at which growth rate is zero, in this case at a density of 108 females.

The regression also provides some information on the variance in

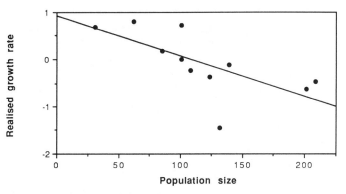

Figure 3.13 Regression of population per capita growth rate, r_t, on female shrew population size for November ($r_N = 0.917 - 0.00848N$, adjusted variance explained $= 0.39$).

growth rate. Environmental stochasticity in our model is represented by observed deviations from the deterministic expectations. The deterministic expectation is represented by the regression line in Figure 3.13. The unexplained variance, 0.61, is due to variability in the environment.

We can estimate the magnitude of environmental variance from the observed deviations around the regression line, the regression residuals. The mean squared deviation from the regression is the expected variability. In Figure 3.13, the average distance, squared, of an observation from the regression line is 0.262. We shall use this as our estimate of the level of variation in the growth rate of the population that results from environmental variability.

3.4.4   Parameter estimation for the Ricker equation

Again, all we have at our disposal are the census data for 12 years. Estimating the parameters for the Ricker equation is as straightforward as it is for the discrete logistic equation. We can fit a linear regression to the data to give values for r and β by rearranging Equation (3.17). Using the values for the shrew female population density over 12 years from the month of November, the best fit for the November data gave

$$\ln\left(\frac{N_{t+1}}{N_t}\right) = r - \beta N_t = 0.94 - 0.0084N_t \ . \tag{3.21}$$

Note that the form of this equation is the same as Equation (3.8) used to estimate the parameters of the logistic. Equation (3.8) works for the logistic because we assume r and K represent the same things in the continuous form (Equation (3.10)) and the discrete form (Equation (3.12)). Even though the estimation procedures are the same for the logistic and the Ricker equations, the parameters play very different roles when

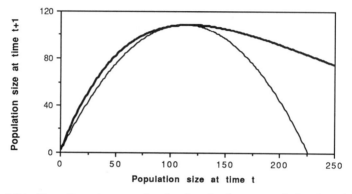

Figure 3.14 Recruitment curves for the female shrew population based on the Ricker function (heavy line) and the discrete logistic function (thin line). The parameters of the Ricker equation are $r = 0.917$ and $\beta = 0.0085$ and of the logistic are $r = 0.917$ and $K = 108$.

substituted into the respective equations. In the Ricker equation, like the logistic, these parameters result in a carrying capacity of 108 females (Figure 3.14).

The recruitment curve for the Ricker function (Equation (3.21)) is quite flat after the inflection point, so the effects of scramble competition on the dynamics of the population will be weak (Figure 3.14). In fact, the dynamics of the population will be close to those expected under contest competition (Figure 3.9). There will be few of the dramatic fluctuations in population size that result from the logistic equation.

3.4.5 Sensitivity analysis

To illustrate what sensitivity analysis means, consider the equation $A = B + 0.00001C$. If we know that the parameters B and C are approximately the same magnitude, then B will contribute much more to the value of A than will the parameter C. It is plain to see that it is much more important to estimate the value of B carefully than it is to estimate the value of C. Change C by 100% and A will change by about 0.00001%. Change B by 100% and A will change by about 100%. Any errors in the estimate of B will have a much greater influence on the value of A than will errors in the estimate of C. In fact, in this case we may decide that C is redundant and omit it from the model altogether. These observations constitute an analysis of the sensitivity of the model. We have evaluated the parameters and determined how sensitive the model is to each of them.

The important thing to estimate in a model is the state variable. In risk assessment for conservation biology, it is usually the population size. In formal terms, sensitivity analysis measures the response of state variables in a model to small variations in the values of parameters (Beck, 1983). It has two purposes. First, it tells us if the results of a model are very sensitive to the estimate of the value of a parameter. If so, we need to concentrate effort on obtaining a value as close the correct one as possible. Second, it tells us if any of the parameters in a model are redundant. If the model's predictions fail to respond, or respond very weakly, to a change in the value of a parameter, we could omit it from the model. In any case there is no need to worry if the estimate of its value is not accurate.

The analytical approach is to calculate the first derivative of the model equations with respect to each of the parameters. A numerical alternative is to measure the relative variation in the state variable(s) that results from a given change in the value of the parameter(s). The sensitivity of a variable, L_i, to the parameter P_j is defined as

$$s_{i,j} = \frac{1}{n} \sum_{t=1}^{n} \frac{dL_{i,t}/L_{i,t}}{dP_j/P_j} \qquad (3.22)$$

where dL is the change (d) in variable L and where the sensitivity is estimated over n time steps (see Beck, 1983). Under this equation, variation of 10% in a parameter that results in a uniform 10% variation in a state variable will give a coefficient of sensitivity of 1. This is the approach we shall use to evaluate the sensitivity of the Ricker and logistic models for the shrew population. More details on sensitivity analysis are provided in the Appendix, Section A.4.

Equation (3.22) measures the proportional deviation that results from a given proportional change in a parameter. We simulated the dynamics of the population over 50 years, increasing each of the parameters in each of the models in turn by 10% (Table 3.1). There are two components to consider in each of these models, the rate of population increase from low densities, and the equilibrium population size (the carrying capacity of the environment). We investigated the sensitivity of each of these components separately. We began each simulation with an initial population size of one female. The effect of changes in parameters on the rate of population increase was calculated by applying Equation (3.22) to the first few time steps of the simulation. The effect of changes in parameters on the carrying capacity was calculated by applying Equation (3.22) to the last time step of the simulation.

Population increase in the logistic model for the shrew population is most sensitive to changes in the instantaneous growth rate, r, while the equilibrium size depends solely on the carrying capacity, K. We could have guessed these results because of the way we formulated the model in the first place. However, the sensitivity the Ricker model is less intuitive. Again, population increase is most sensitive to changes in

Table 3.1 Sensitivity, S_i, of increase in the shrew population size and the equilibrium population size to a 10% increase in each of the parameters and the initial population size, N_0, in the logistic model and the Ricker model. Sensitivity of population increase in the logistic model was calculated using the first ten time steps of the simulation. It was calculated in the Ricker model using the first seven time steps. Sensitivity of the equilibrium population size was calculated in each model using the last time step; all trajectories had stabilized by that time.

Parameter	Population increase		Equilibrium size	
	Logistic equation	*Ricker equation*	*Logistic equation*	*Ricker equation*
β	–	0.31	–	0.92
K	0.28	–	1	–
r	1.14	1.36	0	1
N_0	0.48	0.21	0	0

the instantaneous growth rate, r, but the equilibrium population size is equally sensitive to both r and the parameter β.

The importance of the results of sensitivity analysis depend to some extent on what we want to use the model for. For example, the equilibrium population size is insensitive to initial population size in both models. If we are interested only in the equilibrium population size, we can ignore the effect of initial population size. It becomes important only if we are interested in the size of a population during a period of growth.

One of the most valuable things about sensitivity analysis is that it helps us to understand the equations better. Sometimes, when developing a model for which there are abundant data, the equations we write become long and complex. In such circumstances sensitivity analysis is invaluable in establishing the role and importance of the parameters and the nature of the interrelationships between them.

3.4.6   Predicting the shrew population size

We now have two alternative deterministic models for the dynamics of the shrew populations. We saw in Section 3.2 above how we might add environmental variability to the logistic equation by sampling the growth rate, r, from a random distribution. To model the effect of environmental variability on the population through the Ricker function, we can add a random component

$$N_{t+1} = N_t \, e^{((r+s.y_t)-\beta N_t)} \tag{3.23}$$

where y is a normally distributed random number with a mean of zero and unit variance, representing the variation in the environment at time t, and s is the standard deviation of the variation. This approach is useful

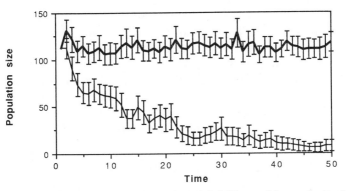

Figure 3.15 Mean female population size and 95% confidence limits for the stochastic models based on the logistic equation (thin line) and the Ricker equation (heavy line). Parameters for these equations are the same as those shown in the legend of Figure 3.14, and the standard deviation in the instantaneous rate of growth for both models was 0.5.

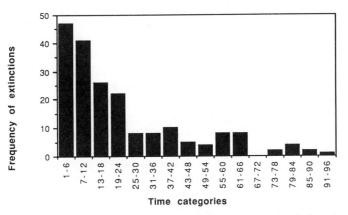

Figure 3.16 Times to extinction (in years) for the shrew population, based on 200 replications of the logistic model incorporating environmental variance. The data are grouped into categories of six adjacent years. Two replicates survived the 100-year period and were omitted from the figure.

for the sake of comparing the results to those of the logistic equation because in both cases we model environmental variability through variation in the population growth rate. It may be equally valid to model variation by adding a random component to the parameter β.

The stochastic logistic model for population growth of the shrew population predicts rapid decline when we use a magnitude in variance equal to that observed in the natural population (Figure 3.15). The mean expected population size after 20 years is less than 50 females, and it declines to nearly zero by about 50 years.

The results of the simulation using the logistic model suggest that the population is very unlikely to survive for 50 years, given the observed variability in growth rates. The mean time to extinction is about 23 years (Figure 3.16) but, like the stochastic analysis in Chapter 2, more than 50% of the replicates became extinct before the mean time. The Ricker model, on the other hand, predicts no important chance of extinction within the next 50 years and the mean population size is expected to remain close to the carrying capacity of 108 females (Figure 3.15).

3.4.7   The time horizon and added risk

The simulations above were made over 50 years and the results suggest that extinction is highly unlikely under the Ricker model. Often, for management purposes, we are interested in impacts on population size over different periods of time. Usually, we do not just want to know the chances that the population will decline. If it does decline, what are the chances of it falling below given densities which, for one reason or another, are considered to be unacceptably small? It may also be important to know the chances and magnitude of population decline given different management strategies.

Quasiextinction risk curves provide a means of resolving all of these questions. Above, we simulated the growth of the shrew population over a 50-year period using the Ricker function to represent the dynamics of the population. From each replication, we recorded the smallest value for population size (Algorithm 3.1). These values were used to generate the quasiextinction curves shown in Figure 3.17. The thin line furthest to the left on the diagram represents the chances that the shrew population will fall below various population sizes (termed threshold population sizes) at least once during the next 50 years.

Assume for the moment that we decide a population size of 40 females is unacceptably small because population sizes below that threshold will suffer important levels of genetic inbreeding, reduced fecundity and loss of genetic heterozygosity. If we go to the x-axis of the figure, to the population size 40, then we can see that there is about a 90% chance that the shrew population will fall below this level at least once within the next 50 years.

The conclusions we draw from these analyses depend on the time horizon over which we make projections. The time horizon is the period over which we run the simulations, 50 years in the case of the shrew and the goose populations. To reduce the time horizon means that we run the simulations over a shorter period. If we reduce the time horizon for the shrew simulations to 20 years, the chance of extinction is greatly reduced. The quasiextinction curve for populations over 20 years is moved to the right of the curve based on simulations over 50 years. This means that the chances of population decline are reduced. The chances of passing below

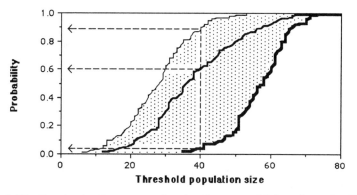

Figure 3.17 Quasiextinction curves for the shrew model based on the Ricker function and using a standard deviation for growth of 0.5. Simulations were over 50 years (thin line) and 20 years (medium line) using a standard deviation for growth of 0.5, and over 50 years using a standard deviation for growth of 0.3 (heavy line). The shaded area represents the reduced risk that results from decreasing the standard deviation of growth in the model from 0.5 to 0.3.

the critical threshold of 40 females is reduced from 90 to 60%. This result makes intuitive sense because if we use fewer years in our simulations there are fewer chances for the population to experience periods of poor environmental conditions.

Assuming the time horizon of 50 years is appropriate, we would like to judge the effect of management alternatives on the shrew population. We could attempt to ameliorate the effects of environmental variation on the population. One way to do this would be to supplement their food. This would effectively smooth out some of the variation experienced by the population, reducing the number of deaths in poor years. We assume that the standard deviation in growth of the population is reduced from 0.5 to 0.3 as a consequence of our actions. If we run the simulation again over 50 years, but with smaller variation, the chances of population decline are reduced. The quasiextinction curve for this scenario is on the far right of Figure 3.17 and the shaded region represents the reduction in risk that results from decreasing the level of variability from 0.5 to 0.3. The chances that the population will fall below the critical threshold of 40 females at least once is less than 5%.

3.4.8   Model predictions and shrew biology

For the analyses using the logistic equation, the expected persistence time is about two decades. Given observed levels of variability in population growth, persistence for more than two decades is unlikely. Using the same levels of variability, the Ricker model predicts that the population will very probably persist for more than five decades.

The deviations of growth rate from the growth expected in the logistic model in November (Figure 3.13) are well explained by temperature and rainfall in October and November, and by the amount of snow in the preceding winter (Burgman *et al.*, 1992). Adverse climatic conditions can dramatically reduce shrew populations: rainfall has an effect on the abundance of soil arthropods, the main source of food for this population in summer, and the depth of snow in winter may be directly related to over-winter survival (P. Vogel, unpublished data; Genoud and Hausser, 1979). If variation in population growth is indeed directly related to variation in the weather, it leads to the conclusion that the population is at risk of extinction after unusually heavy snow in winter, or hot conditions in October and November.

The short expected persistence times are surprising. This population represents one of the larger and more stable populations in suburban areas near Lausanne. The results could be explained if the *a priori* assumption that major roads, building sites and intensive agricultural land are absolute barriers to migration, is invalid. Young individuals of this species disperse from their parental territory in spring and summer (Cantoni and Vogel, 1989). There are records of other, smaller populations in the vicinity that regularly become extinct over winter and are often recolonized by dispersing juveniles in the following spring (Genoud, 1981), but nothing is known about the distances covered or the barriers crossed by migrating animals of the species. Dispersal is known to be a critical factor in the persistence of metapopulations in both natural and urban populations of several other small mammal species (Gliwicz, 1988; Dickman and Doncaster, 1989). Migration may play a critical role in the survival of *Crocidura russula* in suburban areas in western Switzerland and the metapopulation structure and probabilities of successful migration may determine its continued existence in this and similar habitats.

An alternative explanation may lie in our reliance on the logistic model for density dependence. There are at least two ways in which the logistic model could go astray. First, as many as 44% of juvenile female *C. russula* can reach sexual maturity in the same year as their birth and participate in reproduction, giving two effective generations in one calendar year (Besançon, 1984). Besançon suggests that juveniles reproduce in the same year as their birth whenever the population density is low. Second, while the habitat for shrews within the study area is superficially uniform, there may exist a sheltered core area, or 'hot-spot', within which conditions for population growth are favourable even when population densities are low (Goodman, 1987c). Both of these points imply higher growth rates than predicted by the logistic equation and/or smaller variances in growth rates when the population is small.

The most obvious reason for the expected predictions of the logistic is that the underlying dynamics of the population were not well described by it. The fluctuations in population size inherent in the logistic, especially those that result from high population densities, may not eventuate in reality. The predictions of the Ricker model for expected extinction times are much more conservative. This model predicts that the population is unlikely to become extinct within the next 50 years although there is a small chance the population density could fall to as low as 6 females.

It is not possible to say that migration, particular environmental variables, or highly sheltered microhabitats are critical for the persistence of this shrew population. The events leading to extinction are fundamentally stochastic processes, and we have observed only one realization of these events in the Préverenges population. Neither is it possible to say that the Ricker model provides a more reliable prediction for future shrew population sizes than does the logistic. We prefer the Ricker model on the grounds of available autecological information and because the results it provides agree more closely with our preconceived notions of the sorts of risks faced by the population. We can only conclude that the expected persistence time for the population is sensitive to the magnitude of environmental variance and that, if the Ricker model is an adequate summary of its dynamics, we expect it to persist for more than several decades.

3.5 MORE ABOUT UNSTRUCTURED MODELS

3.5.1 Allee effects

Usually, density dependence represents the decline in population growth that occurs when population densities are high, and it results in populations reaching an equilibrium abundance when resources are limiting. The Allee effect operates at the other end of the density spectrum. It is a form of density dependence that occurs when, below some critical level, population growth declines as population density declines. It gets its name from W. C. Allee (Allee *et al.*, 1949) who studied sociality and cooperation in natural populations.

When populations become very small, inbreeding among related individuals becomes unavoidable and may create genetic problems in habitually outbred species (see Chapter 6). The effect may also be observed when reproduction opportunities are diminished in populations so sparse that individuals cannot easily find mates, or when the group ameliorates the environment in some important way. For instance, disjunct populations of Hemlock in Indiana seem to acidify the soil and sequester water in the upper horizons. Both of these processes facilitate hemlock reproduction and such phenomena require locally high densities (Ferson and Burgman, 1990). At low densities, population growth rates will decline because there are too few individuals to be able to modify the environment suitably.

Whenever fewer individuals results in reduced chances of survival and reproduction for those that remain, Allee effects occur. This may lead to a situation in which species are led to an extinction vortex in the sense of Gilpin and Soulé (1986).

Allee effects and primate population dynamics
About half the 200 extant primate species around the world are threatened. It is difficult to suggest any general rules for their management or conservation because they exhibit a very broad range of complex social structures, in addition to occupying a wide variety of habitats. Their organization ranges from individual pairs that mate monogamously to large promiscuous groups. Dobson and Lyles (1989) reviewed the literature on primate demography and used it to develop a set of generic models of the sort we outline in Chapter 4 below.

Group size in primates is the result of interactions between their social system, the distribution and abundance of resources, and the degree of habitat fragmentation (see Chapter 5 below). The ability of individuals to find mates is likely to be dependent on both population density (the Allee effect) and the degree of subdivision of the population into social or habitat fragment groups. Dobson and Lyles' (1989) results illustrate a point that had previously not been perceived as important in the management of primate populations. Species that tend to live in aggregated groups and that exhibit promiscuous matings will establish and maintain themselves at smaller population densities than species with more solitary and monogamous habits. That is, in general, social dysfunction occurs at higher densities in species with a tendency towards a solitary lifestyle.

A specific example of this kind is provided by Robinson (1988) who studied the demography of wedge-capped capuchin monkeys (*Cebus olivaceus*) that occur in the savanna plains of central Venezuela. Groups of the species are found in gallery forests bordering rivers and in shrub woodlands which are near permanent water. An indication of the rate of increase of a group is given by the proportion of juveniles in a group. It seems reasonable to speculate that those with relatively few juveniles are not increasing as quickly as those with many juveniles. The relationship between group size and the percent of juveniles (Figure 3.18) in this species suggests the Allee effect. Small groups have a smaller proportion of juveniles than large groups.

There are many ways in which Allee effects may be included in a model. Dobson and Lyles (1989) give an example of one approach. They

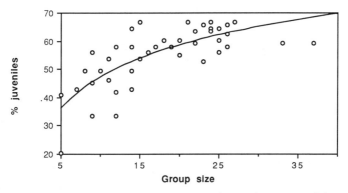

Figure 3.18 The relationship between group size and percent of juveniles in groups of wedge-capped capuchin monkeys in central Venezuela (after Robinson, 1988). Each point is an annual census of one group. The curve is a least squares best fit of the form $y = 9.6 + 37.6 \log(x)$, $r^2 = 0.57$.

assume an underlying statistical distribution for primate party sizes and use it to estimate the probability of finding a mate. Akçakaya and Ferson (1990) introduced a third parameter to the Ricker equation to incorporate Allee effects. They represented the dynamics of the population by

$$N_{t+1} = N_t \, e^{(r - \beta N_t)} \left(\frac{N_t}{A + N_t} \right). \qquad (3.24)$$

In this model, Allee effects are felt continuously, but they increase in magnitude as the population becomes smaller. The strength of the Allee effect in a population may be represented by manipulating the parameter A. The horizontal, broken line in Figure 3.19 represents the threshold at which the population exactly replaces itself from one time step to the next. Below a density of about six individuals, the population will decline to extinction due to the systematic pressure of the Allee effect. When species fall below the critical population size, population growth diminishes and extinction risks may be seriously underestimated if the effect is ignored.

3.5.2 Birth and death revisited

In 1960, Bartlett made the important observation that if the birth rate and the death rate are functions of population size, one can have models to account for demographic stochasticity that mimic density-dependent models of deterministic population growth. MacArthur and Wilson (1967), MacArthur (1972) and Richter-Dyn and Goel (1972) experimented with some of these models. They found that the expected lifetime of a population increases with increasing growth rate, with lower overall birth and death rates, and with higher carrying capacities. Most

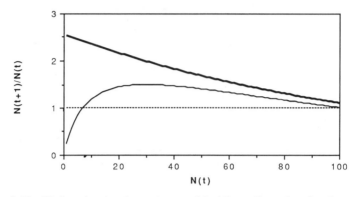

Figure 3.19 Ricker density dependence with Allee effects for the shrew population. The upper, heavy curve represents the expected increase in the shrew population as a function of a range of current population sizes ($N(t)$), using the estimated parameters $r = 0.94$ and $\beta = -0.0084$ without Allee effects ($A = 0$, Equation (3.24)). The lower, thin curve represents the same model for the shrew population, but with Allee effects present ($A = 10$, Equation (3.24)). (After Akçakaya and Ferson, 1990.)

importantly, they found that very long persistence times could be expected for populations with modest population ceilings. This led to the general conclusion that populations in excess of about 20 individuals would not be significantly affected by demographic stochasticity.

Goodman (1987b) used the framework of birth-and-death models to develop an equation for unstructured population models to calculate the probable time to extinction. The innovation in Goodman's model is that it uses functions for mean population growth rate (r) and variance in growth rate that depend on population size, and the model accounts for uncertainty arising from both demographic and environmental sources. To calculate a mean time to extinction, one needs the growth rate of the population and the variance in growth rate, both as a function of N. Burgman *et al.* (1992) applied Goodman's equation to the population of suburban shrews we modelled above.

Goodman (1987a,b) developed numerical examples using his method, exploring the influence of maximum population size, variance in growth rate and the strength of temporal correlations on estimated persistence times. He found that variance in growth rate, particularly that component due to environmental variation, will prove critical in determining the mean extinction time for a small population. Persistence increases only gradually with increasing population ceiling. The mean growth rate will have a small effect on mean persistence times when the population ceiling is small. When the population ceiling is large, there is an appreciable effect of r, especially when r is small. He tested his equation against alternative age-structured formulations and ran simulations with varying degrees of autocorrelation in environmental conditions. Strong autocorrelation in environmental favourableness and the effects of an age-structured population do not greatly affect the predictions of the equation.

3.5.3   Time lags

In the logistic equation in continuous time (Equation (3.10)), the damping effect of density dependence is manifested instantaneously. However, in the real world, there are always time delays resulting from the generation time of the species or the recovery of a resource. Time lags may be included explicitly in continuous time models. May *et al.* (1974) report on the general characteristics of some of them. One of the most flexible is

$$\frac{dN}{dt} = rN_{t-g}\left(1 - \frac{N_{t-w}}{K}\right). \tag{3.25}$$

Here, g is the reproductive time lag (the gestation time or its equivalent) and w is the reaction time lag, the time between changes in the environment and the corresponding change in population growth rates (see Krebs, 1985, pp. 226–8, and references therein). The time lags are related to the migratory, exploratory and reproductive behaviour of individuals in a population. This equation takes into account the fact that the reaction by part of a population in one place to what happened in another place cannot be immediate (Lomnicki, 1988, p. 102). For example, a solitary animal dies in a remote corner of a forest. In the continuous logistic equation, (3.10), the instant an individual dies the resources it commanded are assumed to be available for other individuals. However, it may take time for the event to be discovered by conspecifics and there will be a lag before the resources in its territory contribute to the survival and reproduction of others in the population.

  We do no more than note the continuous time form of Equation (3.25). Continuous time models and the associated analytical methods such as stability analysis are beyond the scope of this book (cf. Nisbet and Gurney, 1982).

  In the discrete time models (Equation (3.12), (3.16)–(3.19)), there is an implicit time lag in response of growth rates to increases in population density, proportional to the length of the time step chosen. This time lag is responsible for the unstable behaviour of the logistic model when growth rates are high. When applying these models, it is important to be careful that the time step chosen is biologically meaningful. The time lag in the model should reflect real time lags in age structure or rescource recovery time.

  It is possible to write equations for populations in discrete time that include an additional time lag component. A useful model in discrete time that includes an explicit time lag was used by Turchin (1990):

$$N_{t+1} = N_t \, e^{(r+\beta N_t + \gamma N_{t-1})} \tag{3.26}$$

and is an extension of the Ricker model (Equation (3.17)). It says that the population size at time $t + 1$ is a non-linear function of both the population size at time t and the population size at time $t - 1$. Turchin applied the model to censuses of 14 forest insect species and found eight cases to exhibit evidence of delayed density dependence and lag-induced oscillations in population size. Density dependence without the time lag could be detected in only one of the 14 species considered in the study. Delayed density dependence is the effect that the population size at time $t - 1$ has on the population at time $t + 1$.

Turchin suggests that it may arise because high population density adversely affects the fecundity of the next generation. For example, Peterson and Black (1988) suggest that a history of crowding may make individuals in a population of sand-flat bivalves more susceptible to subsequent stress. Such a phenomenon will result in a time lag in which the population size at one time has a direct effect on the population at a later time, and the effect may act over several time steps.

3.5.4   When should we accept a model?

One of the animals to which Turchin (1990) applied the three-parameter model (Equation (3.26)) was the larch budmoth (*Zeiraphera diniana*). This insect attacks subalpine larch-cembran pine forests (*Larici pinetum cambrae*) of the European alps at regular nine-yearly intervals. The phenomenon has been observed for many centuries. Census data recorded from the Engadine Valley in south-eastern Switzerland represent one of the best recorded sequences of population densities for the species (Figure 3.20).

Turchin noted that when the best fit is found between these census data and the Ricker function (Equation (3.17)), the resulting model explains almost none of the variation in population densities observed between 1949 and 1986 in the Engadine Valley. When Equation (3.26) is used instead, more than 50% of the variation in population density can be explained. The time-delayed model produces periodic peaks (Figure 3.20) and Turchin concludes that this series of population densities is best explained by assuming delayed density dependence is important in the dynamics of the population.

Turchin's equation generates periodic cycles and so in one sense it is a better representation of budmoth dynamics than the Ricker equation (Figure 3.20). His intention was to explore the pervasiveness of delayed density dependence in a large number of data sets and he was not concerned with the biology underlying the observed population dynamics. However, there is a real danger in equating improved statistical fit

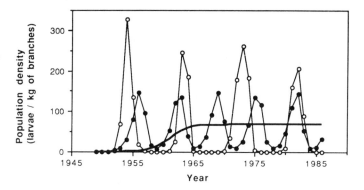

Figure 3.20 Population densities for the larch budmoth in the Engadine Valley (after Baltensweiler and Fischlin, 1988; open circles) and the values for the population generated by the Ricker (heavy line) and Turchin (closed circles) models using parameter estimates made by Turchin (1990).

(explanatory power) with biological understanding and predictive power. There are many things that can cause regular cycles in a population, such as scramble competition together with a relatively high population growth rate, and climatic cycles. The fact that Turchin's model generates cycles is only weak inductive evidence for delayed density dependence in the budmoth population, and is convincing only in those cases where the pattern it predicts closely matches the observed cycles. In the case of the larch budmoth, the period of the theoretical population cycles is too short. Turchin's model generates five cycles in the time that the real population exhibits four.

Turchin did not attempt any predictions. He was intent on exploring the qualitative dynamics of the population censuses. The kinds of analyses we develop in this book differ from the kind represented by Turchin's study in that predictions are central to the role of models in conservation biology. If we were to make predictions about the timing of explosions in larch budmoth populations, Turchin's equation is unlikely to be much help. The Ricker model displays no cycles at all and captures none of the qualitative dynamics of the population. Neither of these models captures the real dynamics of the population and one should be very wary of using either of them to make predictions about future population sizes. The next step would be to modify the way in which the model works, making it a better representation of the important processes that determine population size.

3.5.5 Explanation and understanding

In this section, we develop a hypothetical example to illustrate the difference between statistical explanatory power and predictive power. Our ultimate intention is to make predictions about the risks of decline of a

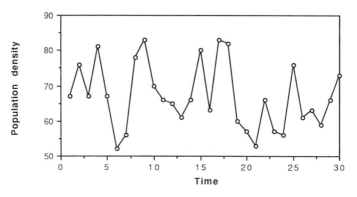

Figure 3.21 A sequence of population densities for the hypothetical species generated using Equation (3.27) and a CV of 30% in the growth rate to represent environmental variability. The carrying capacity of the population is 67.

population, so we go beyond exploration of the kind made by Turchin in his evaluation of forest insect censuses.

There is a chance that all populations are affected in some way by the density of the population at times other than the immediately preceding census. So, why not use Turchin's equation exclusively and forget about the Ricker equation? If delayed density dependence is unimportant in the dynamics of the population, the term representing it in the model, γ, will reflect that fact and will explain only a small amount of the variability in observed population sizes. By including the extra term, we cannot lose. Therefore, won't a model with three parameters (Equation (3.26)) always be a better descriptor of population dynamics than a model with only two parameters (Equation (3.17))?

The short answer is, no. In fact, if you get the equation for the model wrong, the predictions it makes may be misleading, even if it explains more of the variation in observed population sizes than does the correct equation. To illustrate some other dangers of using the wrong model, we shall assume that the Ricker model for a population (Figure 3.20) is true; that is, the population dynamics of the species are described by the equation

$$N_{t+1} = N_t \, e^{(0.54-0.008N_t)}. \tag{3.27}$$

Further, we shall assume that environmental variability affects the population growth rate and that the coefficient of variation in growth rate is 30%. Using Equation (3.27) with random variation in r (see Equation (3.23)), we generated a series of population sizes. This series represents the censuses that might be observed given that the assumptions we made are true. Observed variation in population sizes from year to year results from random variation in environmental conditions together with the

Table 3.2 Measures of fit between a series of population densities generated using the Ricker equation with a coefficient of variation of 30% in growth, and the Ricker and Turchin population equations.

	Ricker		Turchin	
Years of data	Adjusted r^2	Probability	Adjusted r^2	Probability
5	0.05	0.35	0	0.69
10*	0.30	0.06	0.49	0.04
20	0.35	0.004	0.41	0.004
30	0.37	0.0002	0.38	0.0006

* Parameters of the Ricker model: $r = 0.75$, $\beta = 0.011$.
 Parameters of the Turchin model: $r = 1.23$, $\beta = 0.009$, $\gamma = 0.009$.

deterministic variations caused by density-dependent mechanisms within the population.

Now, we shall take the role of a biologist unsure of the true form of density dependence. The only information is the series of points on Figure 3.21 representing population densities recorded at yearly intervals. We know enough about the species to assume that scramble competition for resources is possible at high population densities. We have to choose between the two-parameter Ricker equation, (3.17), and Turchin's three-parameter equation, (3.26).

The first thing we might do is fit the equations statistically to the observed data (Table 3.2). If we equate statistical explanatory power with predictive power, the equation that explains more of the variation in population size will provide a better model for the population's dynamics and we would then use it to make predictions for the population.

Turchin's equation always does better than the Ricker equation in explaining the variation in the population (Table 3.2), as it has an extra parameter that gives it an extra degree of freedom. It is mathematically impossible for it to fit any data set worse than the Ricker equation. If choice were based purely on statistical explanatory power, we would always use Turchin's equation. Better still, we could add more and more parameters until the fit became perfect.

The trouble starts when you want to make predictions because you then rely on the biology underlying the equation. If it is incorrect in the sense that it summarizes poorly the dynamics of the population, the predictions it makes will be poor irrespective of how well it fits the observations so far.

Suppose we had only ten years of the above census data at our disposal and we fitted the two equations. The variance explained by Turchin's equation is substantially higher than that explained by the Ricker equation (Table 3.2) and Turchin's equation results in a statistically significant

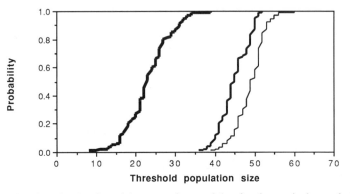

Figure 3.22 Quasiextinction risk curves for models of a theoretical population of insects: the true population risks (thin line), risks calcuated using the estimated Ricker model (medium line) and the estimated Turchin model (heavy line). The true population is governed by the Ricker equation with parameters $r = 0.54$ and $\beta = 0.008$. The parameters for the estimated models are given at the foot of Table 3.2.

fit to the data (at the accepted level of 5%) whereas the Ricker equation does not. On these grounds alone, one would use Turchin's equation.

We are asked to evaluate the risks of population decline within the next 50 years. To do so, we simulate population growth (Algorithm 3.1) using Turchin's equation and our estimates of its parameters (Table 3.2).

The result of this analysis is the risk curve on the far left of Figure 3.22. The chances that the population will fall below a density of 30 are very high. The 'true' risks faced by our theoretical population described by Equation (3.27) are given by the risk curve at the far right of Figure 3.22. The model based on Turchin's equation results in the curve at the far left of Figure 3.22 and it has overestimated the risks faced by the population quite substantially. The risk curve close to the 'true' curve in Figure 3.22 is that based on the Ricker model, using the parameters estimated on ten years of data (Table 3.2). It provides a much better estimate of the risks faced by the population, even though it explains less of the variation in the observed data over the first ten years.

The reason for this is that Turchin's equation can generate oscillations, such as those in Figure 3.20, much more readily than can the Ricker equation. The delayed density-dependence component adds another dimension to the dynamics of the population. As a result, when the population is carried away from the carrying capacity by some chance environmental event, Turchin's equation predicts relatively large variations in population size that are entirely deterministic. The population varies more than in the equivalent Ricker model, leading us to overestimate the risks faced by the population.

This example has not been thoroughly evaluated. A complete test would include many parameter combinations for the 'true' model, each tested with many time series. Nevertheless, it serves to emphasize the difference between explanatory and predictive models. The model that provides the best predictions of population dynamics is not necessarily the model that best fits observed data. The primary motivation in formulating a predictive model should be to capture all important characteristics of the population, and it should be based on an understanding of the biology of the species. At the same time, the model should be as simple as possible, without any unnecessary parameters or relationships that are not based on sound biological reasoning.

3.5.6 The problem of parameter estimation

In this section we shall make a test of our ability to estimate correctly the parameters of a model that includes density dependence. It would be interesting to see if it is possible to reconstruct density-dependence parameters when the underlying mechanisms for density dependence are known. If it is possible, it would give us confidence that model predictions may be correct if we know enough autecology to understand the way in which density dependence works for a population.

It turns out that there is good reason to use a delayed density-dependence model for the larch budmoth. One plausible hypothesis rests on the delayed effects that changes in food quality have on subsequent generations (Baltensweiler and Fischlin, 1988). Turchin's model for the budmoth population from Switzerland is

$$N_{t+1} = N_t \, e^{(1.2-0.0001N_t-0.02N_{t-1})}. \tag{3.28}$$

We used this equation with random variability in the first parameter with a coefficient of variation of 30% to generate sequences of population densities. Based on the data for the first five years of the first simulation, our estimates for the three parameters in Equation (3.28) are $r = 1.58$, $\beta = 0.000415$ and $\gamma = 0.0266$, neither of which is very close to the mark. Using the data from the first 20 years, our estimate for the same three parameters are different: $\beta = 0.0011$ is worse, whereas $r = 1.324$ and $\gamma = 0.0202$ are better.

It appears from these results that as the sequence of numbers in the census becomes longer, our estimates of the parameters will change. Hassell *et al.* (1989) found that density dependence is usually detected in larger samples (longer censuses). To verify this generalization, we repeated the simulation five times over 30 years. From the numbers produced by each simulation, we estimated the original parameters in Equation (3.28). We calculated the average value for each parameter at 5, 10, 20 and 30 years over the five simulations (Figure 3.23). The mean values for the parameters tend to converge towards the 'true' value as the sequence extends from 5 to 30 years. Also, the confidence limits for

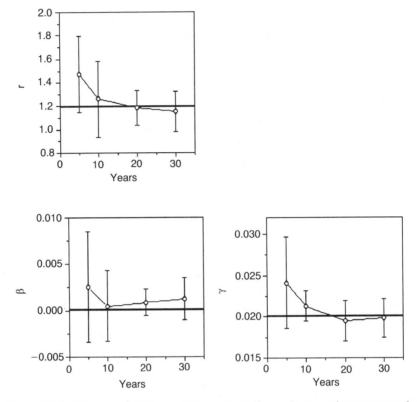

Figure 3.23 Means and 95% confidence limits for estimates of parameters for Turchin's model for larch budworm populations in Switzerland (Equation (3.28)), $r = 1.2$, $\beta = 0.0001$, $\gamma = 0.02$. Averages were calculated from five sequences generated using a CV of 30% in r. The number of years of data from which each estimate was made is shown on the x-axis. The heavy horizontal lines are the values for each parameter.

the estimates become smaller, implying that as the sequence of values in the census becomes longer, we can be more and more confident we are close to the true value.

There are several disturbing things about Figure 3.23. The estimation procedure consistently overestimates the three parameters for the first five or ten years. Furthermore, if the data cover only five or ten years, the estimates we make for the parameters of the model are relatively uncertain. Keep in mind that things are this bad even assuming that the model we use correctly describes all the important components of the dynamics of the population.

Remember that Turchin's time lag model (Equation (3.28)) did much better than the standard Ricker function (Equation (3.27)) in explaining observed variation in the larch budmoth population. The results are qualitatively the same when this experiment is repeated on the two-parameter Ricker model. For the first few years for which data are available, the estimates of these parameters are relatively uncertain and there is a tendency for the estimated value to be larger than the correct value. The results imply for both Turchin's and the Ricker model that censuses covering ten years are not long enough to provide reliable parameter estimates, at least when the coefficient of variation in growth of the population is around 30%.

There are a few circumstances in which the estimation of model parameters may be a little easier. It helps if environmental variability is small, so that it does not mask density-dependent fluctuations. If the census series includes a large range of population sizes, from very sparse populations to dense ones that experience the full effect of density-dependent mechanisms, the problem will be alleviated somewhat. If the population remains close to the carrying capacity for most of the period of record, the parameters will be relatively difficult to estimate.

These results are disheartening because rarely do we have extensive data sets at our disposal. The majority of species for which extensive data sets exist are of some commercial or cultural value to humans. The larch budmoth is extensively studied only because it turns the spectacular Swiss subalpine forests an unpleasant shade of brown in the midst of the tourist season. This led to a study aimed at controlling the peak densities (Baltensweiler and Fischlin, 1988). Members of the Swiss forest service interested in the larch budmoth can rest a little easier because they have more than 40 years' reliable data from which to estimate model parameters. Furthermore, their data econompass a wide range of population densities. They can be confident of being in the right ball park if they are able correctly to model the species population dynamics.

For many of the other models described in Section 3.2 above, the only way to estimate parameters is to use non-linear regression or some other non-linear estimation method. Usually, programs require that you supply the equation and a series of numbers representing census data. The programs will then try to find values for the parameters in the equation that result in the closest fit between the model predictions and the observed population densities. The procedures these programs employ are analogous to the linear regression we used for the equations above. Many attempt to find parameters such that the sums of squared deviations between observed and predicted values are minimized. It is just that there is no easy way of finding these parameter values, so the programs employ sophisticated algorithms.

The tendency of estimation procedures consistently to overestimate the parameters in a non-linear model is not unique to the regression approach we used. Nor is the need for 20 or more years of data before you can be confident that parameters for many models are more or less correct, even in the presence of modest amounts of environmental variability. Ginzburg *et al.* (1990) tested the accuracy of reconstructing the parameters of the Ricker density-dependence model. Even with the advantage of knowing the true mode of density dependence, they showed that with a single data set spanning ten years, the estimation of both parameters might be very wide of the true values. The non-linear estimation procedure they used consistently overestimated both parameters, a result that had been demonstrated before (Walters, 1985; Walters and Ludwig, 1987; Caputi, 1988).

Morris (1990) came to similar conclusions when investigating the adequacy of fitting short-term data from insect populations to unstructured models, and also pointed out that parameter estimates may be sensitive to the fitting procedure used. Biomathematicians such as Schaffer and Kot (1986), May (1989) and Sugihara *et al.* (1990) have only recently begun working on these problems in earnest and it is clear that highly non-linear relationships such as those for scramble competition will require careful sampling and experimentation to unravel.

3.6 SUMMARY

Environmental variability may be incorporated in the exponential model by allowing the growth rate parameter to vary randomly. Variance in the expected size of a population is greater when the mean growth rate of a population is higher, but the probability of quasiextinction falls as the mean growth rate increases. The effect of demographic variability on the probability of extinction depends critically on population size and on the magnitude of mean birth and death rates.

The logistic equation is one of the most widely applied equations for predicting population growth. It is used to model magpie geese in northern Australia. Maximum sustainable yield predictions are substantially lowered when environmental variance is included through random variation in the population growth rate. The logistic in continuous time predicts very different population dynamics to an analogous equation in discrete time.

Several useful unstructured models have been developed in recent decades. The choice of which to use for a particular case should be based on autecological information concerning intraspecific interactions, and not just on the statistical fit of a model to observations of past population sizes. It is most important to have an understanding of the population dynamics and life history of a species, or carry out manipulative experiments on it.

A model for shrews in western Switzerland is developed using both the logistic and the Ricker equations, and a general algorithm for density-dependent population growth that incorporates environmental and demographic stochasticity. Sensitivity analysis assists us to understand the relative importance of different parameters for the dynamics of a population. The logistic and Richer equations make very different predictions for the size of the shrew population because of differences in the strengths of their density-dependence mechanisms.

Equations in discrete time have an implicit time lag proportional to the time step of the model. Explicit time lags may be built into the equations for population growth. Estimation of parameters for even the simplest models can be a nightmare, especially when the sequence of observed population sizes is less than 10 or 15 years. For species for which no long-term census records exist and for which density-dependent mechanisms are important, the results of any modelling exercise should be treated with extreme caution.

4 Structured populations

In previous chapters we have represented a population with a single number, N, which varies over time. In this chapter, we introduce a new class of population models based on matrices that have a very important role in conservation studies. These matrix models are designed to capture the internal structure of a population, using a vector of numbers to represent the abundances for each of the components of the population. Although they require considerably more data to parametrize, such models can yield substantially more detailed forecasts about the future of a population. In cases where the composition of the population has been disturbed, structured models may be the only way to predict the consequence.

In this chapter we shall describe basic matrix models of age- and stage-structured population dynamics as well as a few elaborations that have been used or suggested by biologists. In reviewing the standard development of matrix models, we point out the analyses and results that are especially relevant in conservation biology. As a part of our general thesis that the most useful models for conservation studies are short term and probabilistic, we also argue that the needs of conservation biology demand a focus somewhat shifted from the traditional one that has shaped matrix modelling of population dynamics.

4.1 AGE STRUCTURE

The models that have been discussed so far have made a very strong assumption about the populations they represent. They have supposed that differences among individuals make no difference to the overall dynamics of the population. This might be true if all individuals in a population were identical or if whatever differences they have were to somehow cancel each other out at the level of the population. It is clear, however, that there are important differences among individuals, both with respect to how many offspring they produce and with respect to their chances of mortality. As we shall see, these differences can have profound influences on the behaviour of the population as a whole. Therefore it may be necessary to have a more detailed model that can represent these important differences among individuals that compose the population. By incorporating such detail, we can do without some of the assumptions that the simpler versions require but at the cost of greater complexity. In principle, we could model every single individual separately and, perhaps surprisingly, some workers have suggested doing this (see the discussion by Levins, 1966). Short of this extreme, we can partition the population into classes which differ in the characteristics important to the model. Individuals within a class would be similar to one another, at least enough so that grouping them together results in a reasonable approximation.

Age is one of the most striking differences among individuals that affects their demographic characteristics. For example, very young plants and animals generally do not reproduce. In fact for many species, differences in the number of offspring to different individuals is largely determined by age. There are also other differences related to age that can have important impacts on the population's dynamics. Dispersal is often confined to certain age classes. For instance, in some small mammal species such as the shrew modelled in Chapter 3, juveniles disperse; in plants, the youngest age class, the seeds, are the dispersers. Survivorship is also rather closely tied to age. In species such as whales and gorillas, infant mortality can be high, but otherwise, mortality is low until late in life. Tree species that produce many seeds generally experience a great deal of mortality among those seeds; survival rates for established trees is vastly higher. Even when there are not strong differences among age classes in mortality, the fact that individuals age through time results in continuous attrition from a cohort and a very uneven distribution of individuals among ages.

To represent the distribution of ages within a population, we use a vector of numbers

$$\mathbf{n}(t) = \begin{bmatrix} n_0(t) \\ n_1(t) \\ n_2(t) \\ . \\ . \\ . \\ n_\omega(t) \end{bmatrix}, \tag{4.1}$$

Where $n_x(t)$ denotes the number of individuals of age x at time t. For convenience, t is shown in parentheses rather than as a subscript. The maximum age (or the last age class included in the model) has the subscript ω. Notice that we use a lower case letter to refer to the abundance in an age class; this permits us to retain the capital letter N for the total population size. Age and time may have any convenient units. In our discussion below, we shall sometimes use the word 'year' in an example. Of course there is nothing special about this particular time step. Age structure can be defined in years, months, days, decades or any units that may be convenient. However, age and time must have the *same* units for the age-structured models we shall first discuss in this chapter. In this case, if individuals are grouped into year-long age classes, the time step will also be one year.

Whenever working with age structure, it is important to keep in mind exactly how the ages of individuals are determined. For instance, Americans say that an infant is one year old on the first anniversary of its birth. Before that it would be a 'zero-year old'. Horses in America, on the other hand, are called yearlings between January first of the year after the year in which they were foaled and the next January first. For some species, data are reported with indices starting at one, rather than zero. Any of these methods is acceptable, but this can make a difference to how the various equations are written and it is therefore important to understand what system is being used.

4.1.1 Survivorship and survival rate

The characteristics of several age classes can be described by schedules of age-specific demographic statistics. Schedules pertaining to survival, mortality and longevity are often grouped under the heading of 'life tables' and include a variety of related measures. The most important we shall discuss are the survivorship and survival rate schedules. Survivorship is denoted by the symbol l_x and is the proportion of individuals born who survive to age x. The zeroth survivorship l_0 is defined to be 1; thereafter the numbers get smaller and smaller until the maximum age class, after which of course the survivorship is zero. Life tables can be determined by watching a cohort of newly born individuals as they age, and recording when each individual dies. For a large enough and statistically representative sample, such data can be used to compute an estimate of survivorship.

The survival rate, p_x, is the proportion of those alive at age x who will survive to be alive at age $x + 1$. The survival rate of age x is the quotient l_{x+1}/l_x, and obviously this number has to be between 0 and 1.

The study of life tables originated in the demography of humans and actuarial science. Deevey (1947) reviewed their use in population ecology of non-human species. The life histories of species differ in where mortality is concentrated in the life cycle. Figure 4.1 shows (continuous versions) of the theoretical variety of survivorship functions. The classification of survivorship curves into these types dates back to Pearl and Miner (1935) and Szabó (1931). Type I survivorship results when most individuals live into old age and senesce in a relatively short period of time. The life history of humans is often given as an example tending to this ideal. Type III, in contrast, results from large mortality among juveniles after which the small fraction continue with only slight additional mortality for the rest of the life span. Maple trees approximately exemplify this life history. Continuous mortality of a constant fraction of the population yields the intermediate Type II survivorship curve, for which songbirds are an example. This curve is a straight line when the ordinate is plotted on a log scale.

4.1.2 Maternity rate

Another important demographic schedule is known as the age-specific maternity rate.* This is the number of offspring produced per unit of time per individual of any given age. It is denoted by m_x where, again, the subscript is age. It is often expressed in terms of females only, i.e. as daughters per mother. Like survivorship, the maternity schedule is defined as a continuous function of age. However, it is usually reported as a discrete table of values for groups of ages. Thus, we are given an average number for all those individuals between x and $x + 1$ time units old. Nearly as common though is a schedule where the average is for individuals between $x - \frac{1}{2}$ and $x + \frac{1}{2}$ time units old. The models we discuss below are sensitive to such a difference and care should be taken to discover exactly how the schedule is being defined.

---

* The expressions maternity, fecundity, fertility, natality, and birth rate are all used by various authors to refer to this schedule. In some discussions, these terms take on different shades of meaning, referring to the potential or realized reproduction, successful births, or surviving offspring. But in the current literature the expressions are sometimes used interchangeably. Whenever any of them is used, it is prudent to specify exactly what is meant, including a description of how the quantity is measured empirically.

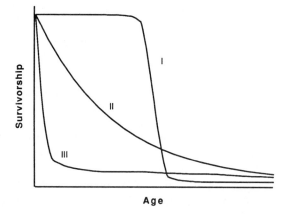

Figure 4.1 Stereotypical survivorship curves (Deevey, 1947; adapted from Pearl and Miner, 1935). See text for an explanation.

4.1.3 Distribution of abundance

Perhaps the most basic schedule in demography (but one that is nevertheless routinely omitted from reports in the literature) is the age distribution. As we said before, this vector is symbolized by n_x. It is conventional in animal demography to ignore the males in a population and express abundances (as well as reproduction) in terms of females (and daughters per mother). Although this coding of the data can be inconvenient at times, it simplifies some calculations. It cannot be done for monoecious species of course, for which minor redefinition of some equations is necessary.

It is important to determine how abundance was measured. For many species this is not always a trivial matter. For instance, in clonal species, are separate ramets or entire genets counted? When are vegetative offshoots counted as distinct individuals? Are there size classes that could not be counted because of limits on observations? Sometimes abundance is reported in units other than demographic individuals, such as biomass. Naturally, the various demographic schedules need to be in compatible units to be useful in the models we describe below. Even then, models with units other than individuals or individuals per unit area typically assume that the population is in a stable age distribution which may or may not be the case. We discuss this point below.

4.1.4 Dispersal

Although it is not a traditional concept, one can also imagine another schedule comparable to survivorship or the maternity schedule which describes the chance, as a function of age, that an individual will disperse from its population. For species in which individuals of particular ages are more likely to disperse than others, this schedule will be an important component of the information necessary for complete modelling of the population.

4.1.5   Variation and uncertainty in demographic schedules

These various schedules summarize the demographic characteristics of a population. It is important to realize, however, that these schedules are not written in stone. A schedule calculated for one population is not necessarily representative of the entire species. Most often a schedule describes a particular population at a particular time and location and under a given set of environmental conditions. It is nearly always the case that the demographic properties of a population vary from place to place in the landscape and over time as a consequence of random fluctuation in the environment and the chance occurrences of genetic change that determine the physiologies and behaviours of individuals in the population. Thus the actual schedules will vary over time in unpredictable ways. The demographic schedules may depend on the population's size and structure. For instance, juvenile dispersal may increase when density rises and home ranges become constricted. If the schedules were calculated using data from a population that was never dense enough to induce this response, the schedules will not reflect it, nor be able to predict it.

Demographic uncertainty in capuchin monkeys
Robinson (1988) studied the dynamics and demography of several wedge-capped capuchin monkey (*Cebus olivaceus*) groups that visited a relatively undisturbed patch of forest in central Venezuela. The results of the study provided an example of Allee effects in Chapter 3. Robinson also found considerable variation in a number of demographic parameters such as the age at which females first give birth (Figure 4.2), and other parameters including the age at which males migrate.

   There is no way to predict these parameters exactly for a monkey group. For example, it is not possible to say that females in these groups begin reproduction at age 6, 7, 8 or 9. We can only say it is usually somewhere between age 5 and 10.

Even if demographic schedules could be specified for a single, well-defined population living in a constant environment and harbouring no density feedback effects, they are not perfect descriptions. As for histograms in general, the way the partition is made out of the variation among ages can influence the overall picture depicted by the schedule.

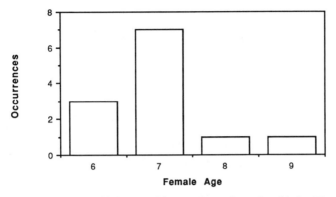

Figure 4.2 The age at which capuchin monkeys first give birth. The data are from several groups observed in central Venezuela by Robinson (1988).

Widening or narrowing the age classes or even shifting the divisions by half a time unit can sometimes have a surprisingly large effect on the schedules and therefore on any conclusions drawn from them, especially when sample sizes are small. And like all empirical observations, the data used in calculating the demographic schedules are subject to measurement error. The modelling methodology for conservation biology should recognize the factors that make demographic schedules unreliable. It should try to propagate the inherent uncertainty in a risk-analytic framework. These aspects of structured models will be developed below, but first we shall outline the use of age structure in building deterministic models for population growth.

4.2 THE LESLIE MATRIX

How can we use all these parameters to project the future growth of a population with age structure? As an example, we shall use a model with four age classes and a yearly time step in which we determine age as we do for humans (that is, you are zero years old until one year after your birth, etc.). In order to construct the projection equations, we need to make some assumptions about the population and its dynamics. We assume that there is no migration or dispersal either into or out of the population. We also assume that reproduction occurs in a fairly short breeding season during the year and that the census is made just after reproduction. Finally, we assume that four is the last reproductive age and that individuals die immediately after they breed at this age.

Since there is no immigration, it is clear that the three year olds next year are just those two year olds this year that survive for another year. We said that the parameter p_2 gives the probability that two year olds survive to be three, so we can write

$$n_3(t + 1) = p_2 n_2(t)$$

Likewise, a similar equation can be written for all age classes greater than zero. We can summarize them all with the expression

$$n_{x+1}(t + 1) = p_x n_x(t) \quad \text{for all } x > 0. \tag{4.2}$$

Now we only need to figure out how to compute the abundance in next year's zeroth age class. The total number of zero year olds next year will be the sum of all the offspring produced by all the individuals who are alive at the next breeding season. Assuming zero year olds themselves do not reproduce in the same breeding season they are born, then we just multiply the abundances in the various age classes by the age-specific maternity rates and add them up:

$$n_0(t + 1) = m_1 n_1(t + 1) + m_2 n_2(t + 1) + m_3 n_3(t + 1) + m_4 n_4(t + 1). \tag{4.3}$$

Notice that the time is $t + 1$ in the parentheses on the right-hand side of the equation. This is because we assumed that the census immediately follows reproduction. Combining Equation (4.3) with the previous survival equations, we see that

$$n_0(t + 1) = m_1 p_0 n_0(t) + m_2 p_1 n_1(t) + m_3 p_2 n_2(t) + m_4 p_3 n_3(t) \tag{4.4}$$

which expresses the fact that, of the individuals alive now, only those who survive the coming year to be alive during the breeding season will get to contribute progeny to the population. Pulling everything together, we can write this equation for the population structure next year

$$\mathbf{n}(t + 1) = \begin{bmatrix} \sum_{x=0}^{3} m_{x+1} p_x n_x(t) \\ p_0 n_0(t) \\ p_1 n_1(t) \\ p_2 n_2(t) \end{bmatrix} \tag{4.5}$$

which encapsulates all our assumptions about the population's structure and dynamics. If you remember how to multiply matrices, it is easy to see that this equation is identical to

$$\begin{bmatrix} n_0(t + 1) \\ n_1(t + 1) \\ n_2(t + 1) \\ n_3(t + 1) \end{bmatrix} = \begin{bmatrix} m_1 p_0 & m_2 p_1 & m_3 p_2 & m_4 p_3 \\ p_0 & 0 & 0 & 0 \\ 0 & p_1 & 0 & 0 \\ 0 & 0 & p_2 & 0 \end{bmatrix} \begin{bmatrix} n_0(t) \\ n_1(t) \\ n_2(t) \\ n_3(t) \end{bmatrix} \tag{4.6}$$

which has the advantage of isolating the vector of this year's abundances on the right as a factor. The only reason this is an advantage is that we can then collapse it to its unsubscripted form $\mathbf{n}(t)$. In doing so, we also collapse the matrix to a symbol $\mathbf{L}$ and say, quite concisely,

$$\mathbf{n}(t + 1) = \mathbf{Ln}(t). \qquad (4.7)$$

This elegant formulation was suggested by independent workers in the 1940s. Although Bernadelli (1941) and Lewis (1942) published the essential idea before him, Leslie (1945) is usually credited with this matrix approach to population projection.

Notice that we did not bother to include four year olds in the vector of age structure. Since we assumed that they die immediately after they breed, at the time of the census these individuals are already dead, thus the abundance in that age class would always be zero. Of course it is not always true that individuals immediately die after they are no longer reproductive. It is possible to follow the post-reproductives by adding rows at the bottom and columns at the left of the Leslie matrix. The new elements on the top row will contain only zeros since these age classes do not reproduce. The new elements on the subdiagonal will determine how the post-reproductives age through time. Be aware, however, that keeping track of post-reproductives in this way makes the matrix reducible (Csetenyi and Logofet, 1989; Caswell, 1989). This is no problem if the dynamics are to be studied by simulation but it can introduce complications for the analytical approaches discussed below that evaluate the asymptotic (very long term) characteristics of the models.

If we assume that the life table and fertility parameters are constant through time (remembering that this is a big assumption), then we can project the population to any point in the future. For instance, if the current population is denoted by $\mathbf{n}(0)$, the projected population distribution for the next time step is $\mathbf{Ln}(0)$. The projection for the time period after that is $\mathbf{L(Ln}(0))$. And naturally, the expectation for t time periods from now is $\mathbf{L}^t\mathbf{n}(0)$. Treating the Leslie matrix as a projection matrix to make these kinds of forecasts about the future of the population is obviously the key to whatever utility the Leslie matrix possesses. In the sections that follow, we shall discuss ways to make these projections without the dubious assumption that the elements of the Leslie matrix never change.

4.2.1 Stable age distribution

If a hypothetical population were to grow according to a constant projection matrix, i.e. one whose elements are unchanging over time, would the abundances asymptotically converge to some distribution? Lotka (1924) established the existence of what he called the 'stable distribution' for population growth of animals. His work was for the continuous case, but it generalizes for the discrete case of a matrix formulation. If the elements remain fixed through time, most Leslie matrices will cause any initial abundance distribution to tend to the stable distribution in the long term. This stable distribution is defined in terms of relative (rather than absolute) abundances and therefore tells the proportion of one year olds to two year olds, and so forth. Once the population has reached its stable form, then, no matter whether the absolute magnitude of abundances is increasing, decreasing or stationary, the shape of the distribution will not change over time. That is, the relative abundances of individuals in each age class will remain constant. You can explore the convergence to the stable age distribution with the following algorithm.

Algorithm 4.1 *Convergence to stable distribution*

1. Fix the parameters of a Leslie matrix **L**.
2. Set the time index t to zero.
3. Initialize the abundance in each age class by setting $n_x(0) = 1/\omega$.
4. Project the abundance vector to the next time step by $\mathbf{n}(t + 1) = \mathbf{L}\mathbf{n}(t)$.
5. Find the maximum absolute difference

$$M = \max_{x} \left| \frac{n_x(t + 1)}{N(t + 1)} - \frac{n_x t}{N(t)} \right|$$

   where $N(t) = \sum_x n_x(t)$.
6. Increment t.
7. Repeat steps 4 to 6 until the distribution converges, say, until $M < 10^{-6} N(t)$ or $t > 1000$.
8. As long as M is sufficiently small, the vector $\mathbf{n}(t)$ is the stable distribution. Scale it to $\mathbf{u} = \mathbf{n}(t)/N(t)$.

Different initial distributions will generally take different amounts of time to converge. You can vary the initial distribution in step 3 of Algorithm 4.1 to see that where the population starts (as long as it is not extinct) will make no difference to the eventuality of convergence. Notice that the stopping condition in step 7 will terminate the algorithm after 1000 iterations. This is important because there are some Leslie matrices for which a population will not necessarily ever settle down to the stable distributions (Bernadelli, 1941; Caswell, 1989, p. 63). For example, if there is only one reproductive age class, the population can cycle forever about its stable distribution even in the absence of any external disturbance. On the other hand, if there are at least two adjacent age classes with non-zero fecundity, then all the oscillations eventually fade and the population will approach the stable distribution. Permanent cycles caused by the structure of the Leslie matrix are usually discounted as biologically unreasonable, although it is conceivable that some species naturally cycle with massive reproduction in what are called mast years (e.g. Janzen, 1976).

It is fairly common in demographic analyses to assume that a population is at its stable distribution, but this presumes that no environmental or anthropogenic perturbations have altered the abundances n_x in the recent past and that the Leslie matrix has been constant for some time. This is a very strong assumption. We suspect that is not tenable in most cases of concern in conservation biology, where disturbance, changing conditions and fluctuating environments are the norm. The stable age distribution is a purely deterministic idea that requires asymptotic time for expression. It is important to remember that actual populations may never approach stable distribution since continual environmental perturbations buffet both the distribution of abundances and the Leslie matrix itself. Although the uses for the stable age distribution are somewhat limited in conservation biology, the notion so pervades the demography literature that it remains an important concept.

4.2.2   Eigenanalysis and the population growth rate

Age-structured population growth has traditionally been characterized by the dominant eigenvalue of the Leslie matrix. An eigenvalue is a scalar number, λ, that satisfies the equation

$$\lambda \mathbf{u} = \mathbf{L}\mathbf{u} \tag{4.8}$$

for some non-zero vector $\mathbf{u}$, which is called an eigenvector. The largest such eigenvalue for a matrix is called the dominant eigenvalue. In the case of a Leslie matrix, the eigenvector associated with the dominant eigenvalue is none other than the stable age distribution. The above equation says that a population at its stable distribution $\mathbf{u}$ will act the same whether it is premultiplied by the Leslie matrix or just (scalar) multiplied by the number λ. Thus $\mathbf{u}$ does what we said the stable distribution would do: it retains its shape and merely changes its magnitude from time step to time step. How much its magnitude changes is given by the eigenvalue λ.

But notice that this change in magnitude is precisely what we mean by population growth. It is the multiplier that tells how the population is rescaled for each time step. If it is less than 1, the population is decreasing. The consequence of such a decline would be eventual extinction. If λ is greater than 1, the population is increasing and would in principle grow geometrically to infinity. Only if it is 1 (by a delicate balance of the values of the elements within the matrix) would the population be stationary. The dominant eigenvalue can be estimated in the same way as the stable distribution.

Algorithm 4.2   Estimating λ

1. Fix the parameters of a Leslie matrix $\mathbf{L}$.
2. Use Algorithm 4.1 to converge to the stable distribution.
3. Once the distribution converges, λ can be estimated as the quotient of total abundances in successive time steps, $N(t + 1)/N(t)$.

The dominant eigenvalue λ measures the asymptotic growth rate of the population; that is, it tells how the population would be changing if the parameters in the model were never to vary and could be fixed for an infinite length of time. λ is often called the finite rate of population increase. Expressed on a logarithmic scale, $\ln(\lambda)$, it is an analogue of the growth rate r for geometric population growth we used in Chapter 3. Notice, however, that population growth at any particular time step is not always given by the dominant eigenvalue. In fact, it is only when the population is at its stable distribution that the population's overall growth is measured by λ. When the initial distribution is different from the stable form, the abundances at the next time step must be computed by working out the matrix multiplication between the Leslie matrix and the current age distribution.

An eigenvalue is sometimes called a characteristic or latent root of the matrix. Depending on the matrix, there may be many eigenvalues but at least one always exists. Subdominant eigenvalues play an important role in the transient behaviour of the population. For instance, how fast a population will approach its stable distribution is influenced not only by the initial distribution but by the ratio of the magnitudes of the dominant and largest subdominant eigenvalue. See Caswell (1989, pp. 68ff) for a discussion. Complexity can arise when several eigenvalues have the same magnitude. In such a case they are called 'repeated' or degenerate eigenvalues. When the dominant eigenvalue is repeated, the cycling described above can be the asymptotic behaviour. This circumstance is rare in the sense that small changes to the elements of the matrix will generally separate the eigenvalues.

Computer software that supports matrices should have routines for computing the eigenvectors of a matrix. Note, however, that the most common eigenvector routines assume that the input matrix is symmetric. Since the Leslie matrix is not symmetric, such routines cannot be used to compute the stable distribution.

4.2.3 Initial distribution

Most of the models that have been offered as descriptions of natural populations predict that the asymptotic population growth rate is independent of the starting age structure. This property, called ergodicity, means that if one waits long enough, the initial distribution of individuals among the classes, no matter how distorted, will eventually be forgotten as the population approaches its stable distribution and asymptotic growth rate. Lotka (1924) showed this for the deterministic continuous case, and it is true – albeit in a weaker sense – for most stochastic matrix models of population growth as well (Cohen, 1979; Tuljapurkar, 1990). We can summarize the theoretical findings in the following way:

1. In a constant environment (i.e. under a constant projection matrix), almost any initial population structure tends to the stable distribution.
2. If two or more populations start from different initial distributions but are subjected to the same mortality and fertility schedules, then, even though these schedules may change over time, the structures of the populations will converge.
3. In a stationary environment (matrix elements are drawn from stationary statistical distributions), the probability distribution of the population tends to a stationary distribution.

Several authors have celebrated ergodicity as a 'liberation' from initial conditions. While certain theoretical deductions may be made easier by ignoring the initial distribution of abundances, it is not the case that initial conditions do not matter. The initial distribution often makes a substantial difference, especially for conservation and management questions.

Suppose that the Leslie matrix for a particular population is

$$\begin{bmatrix} 0 & 10 & 40 \\ 0.05 & 0 & 0 \\ 0 & 0.01 & 0 \end{bmatrix}$$

which is perhaps typical of many species with Type III survivorship curves (Figure 4.1). The dominant eigenvalue of this matrix is $\lambda = 0.72632$. If the initial population size is 100 individuals, what will be the population size after one time step? If your answer is 72, you have fallen prey to the fallacy of λ. (If your answer is 73 or 72.632, you have also forgotten that individuals do not come in fractions.) Determining the correct answer, of course, requires more information that just the total population size. For instance, if the 100 individuals were composed entirely of juveniles in the zeroth age class, then the population after one time step would be 5, since survival rate of juveniles is 0.05 and they produce no offspring. On the other hand, if all 100 individuals were in the oldest age class, then the subsequent population size would be 4000, all of which are offspring of the 100 individuals (who are now dead or otherwise out of the picture). Different distributions of the initial 100 individuals would yield other values between 5 and 4000. From a starting population size of 100 individuals, the possible population size after one time step ranges from 5 to 4000, depending on how the individuals are divided among the three classes. Thus, ignorance of the initial distribution may induce an almost thousand-fold difference in the estimates of population size after only a single time step.

The example above is important for conservation biology because precipitous declines in population size may occur through anomalies in the initial distribution. There are hazards associated with small population sizes that may come into play at such low abundances. These include inbreeding depression, Allee effects, and the increased importance of unlikely events such as disease or increases in predator abundance. If the population falls to a low abundance because of an unbalanced initial age distribution, it may suffer increased risk of extinction from these hazards. In this case, it makes little sense to plan a conservation strategy based on an asymptotic projection of population growth when the near-term chances of extinction are so great that the population may not survive the interim.

It is only when the population is at its stable age distribution (which in this case is 62, 30, 8) that multiplying the abundance vector by the Leslie matrix is the same as multiplying it by the dominant eigenvalue. This makes the stable age distribution a special configuration. Although there is a tendency for a population in a constant environment to approach its stable distribution, variation in the vital rates caused by environmental fluctuations and long-term trends can change the stable distribution. And uneven perturbation to the abundances in the various age classes buffets a population away from its stable distribution. If one can assume that a population is at its stable distribution, then the dominant eigenvalue can be used to estimate future abundance. In general, however, when a population is managed or highly disturbed, the assumption that it is at, or even very near, its stable distribution is often untenable. This is a serious matter that is often neglected. Since many of the species of greatest concern in conservation biology are disequilibrial in one way or another, it is obviously important to be careful when making the assumption that a species is at its stable distribution.

Can we somehow compute the *expected* future population size? We could assume that all possible initial distributions are equally likely. This allows us to estimate the expected population size using a Monte Carlo approach. If the individuals were randomly partitioned into the three classes, the average population size after applying the matrix would be nearly 2000 individuals after one unit of time. This is a vast difference from the sharp decline to 72 individuals obtained by multiplying by the dominant eigenvalue. It is clear that a population far from its stable distribution can yield a radically different growth trajectory than that predicted by the dominant eigenvalue.

The average of 2000 depends on the assumption that all possible starting distributions are equiprobable. If we know nothing else about the population this may be the best assumption to make. Only rarely do we know nothing whatever about a population. In fact, often the most reliable demographic information is on population size and structure. It is fortunate that, once one has constructed a matrix model of population growth, using the abundance distribution in making population forecasts is exceedingly straightforward. Multiplying the observed age distribution by the Leslie matrix gives the best short-term prediction of population growth.

This example reveals the importance of specifying how population size is distributed among various classes. In the near term, the details of the initial distribution can make a substantial difference to the population's future. The example also illustrates how the dominant eigenvalue, λ, can be misleading in short-term projections.

4.2.4 Variability

Many researchers have suggested using the dominant eigenvalue of the projection matrix as a measure of the health of a population. After all, if λ is too small, won't the population become extinct? And isn't it true that the higher its value, the more robust – and safe from extinction – the population is? We argue that the final answer to these questions is 'no'. The reason for this is that making long-term projections based on estimates of current vital rates does not incorporate variability in the vital rates that will occur over time. Furthermore, using the quantity λ as a descriptor of current demography is intimately intertwined with the idea that it is meaningful to project the present abundance distribution infinitely far into the future by assuming the projection matrix does not change over time.

As we have already emphasized, vital rates vary over time. They change for a variety of reasons and in reaction to innumerable factors in the environment. For instance, reproductive, survival and dispersal rates show random fluctuation from season to season induced by environmental factors such as weather. This fluctuation is an inescapable aspect of virtually all real populations, whether they are managed or not. Vital rates sometimes exhibit trends or shifts in magnitude in response to the influence of habitat loss, ecological succession, climate change, or extinction of other species. The influence of these driving forces is often difficult or impossible to foretell before they begin to act. Density dependence, which is a feedback of a population's size on its vital rates, may also occur in many species. In general, density-dependence relations are only poorly understood. Clearly, the variability in vital rates from any of these sources can have a very large impact on what the likely future of the population is. Sections 4.4 and 4.7 will describe ways to incorporate information about the variability in vital rates into models useful in conservation biology. The following example illustrates why it is important to include such variability in a realistic model.

The effect of random variation in vital rates
Boyce (1977) studied the consequence of introducing stochastic fluctuation into the parameters of the Leslie matrix. Instead of assuming that the vital rates were fixed over time, he let them vary randomly from time step to time step, reflecting what they seem to do in nature. Using a model with three age classes

$$\begin{bmatrix} n_0(t + 1) \\ n_1(t + 1) \\ n_2(t + 1) \end{bmatrix} = \begin{bmatrix} 0 & 1.5 & 2.2 \\ p_0 & 0 & 0 \\ 0 & 0.25 & 0 \end{bmatrix} \begin{bmatrix} n_0(t) \\ n_1(t) \\ n_2(t) \end{bmatrix}$$

he simulated the growth of a population whose juvenile survival rate p_0 varied according to a normal distribution with a mean of 0.5 and a given standard deviation. He found that noise in this or, indeed, in any of the vital rates in the matrix systematically reduces the resulting population growth (Figure 4.3). The dominant eigenvalue for the mean projection matrix (i.e. when $p_0 = 0.5$) is 1.01095, which seems to imply a growing population. Boyce found that increasing the noise in a vital rate can cause a population to reverse this tendency, even to the result of population extinction.

The lesson of this observation is that variance in the vital rates can be of singular importance in any demographic study. In much of conservation biology, deterministic and asymptotic analyses are emphasized. It is clear, however, that ignoring the natural variation in demography can lead to a forecast that overlooks impending peril for the population.

What may be as important as Boyce's observation is that the individual trajectories that result from a stochastic simulation can be exceedingly variable. Figure 4.4 depicts several realizations that were obtained with $\sigma = 0.15$. Notice that the ordinate ranges from zero to 20000. One of the trajectories shown became extinct at about the 150th time step. The distribution of survival rates (inset) does not seem unrealistically broad, and would be typical of many endangered species. Yet the resulting population trajectories are, in a phrase, all over the map. Even for fairly small standard deviations in the vital rates, the resulting population fluctuation can be remarkably large.

Any of these realizations could be the *actual* future growth of the real population. Notice that none of the trajectories is at all like the smooth regression curves depicted in the previous figure. Even if the most likely future behaviour of the population can be predicted without bias, the temporal variation within any particular trajectory can be drastic (Ferson et al., 1989). These fluctuations are likely to be important to the conservation effort. For instance, if the population were to fall to a very low level because of chance variation in its vital rates, it could suffer a local extinction or some kind of genetic bottleneck that has long-term consequences for the population. In this way, such fluctuations may pre-empt any predicted asymptotic behaviour of the population.

This example demonstrates how misleading – even in the best of circumstances – the dominant eigenvalue can be as a predictor of future population growth. If the elements in the projection matrix fluctuate randomly through time, the practical utility of λ becomes quite limited. The dominant eigenvalue computed from any of the matrices observed over a single time step will obviously be a poor guess of the aggregate population growth over time. Furthermore, the general theory of stochastic population growth shows that even though the average population will asymptotically grow according to the dominant eigenvalue of the average projection matrix, the most probable trajectories will be below this average (Tuljapurkar, 1990, pp. 43 and 28). This is because the distribution of trajectories is strongly skewed. A few outliers will reach extremely large abundances, but most of the trajectories will have considerably lower abundances. The larger the variances in the elements, the more strongly skewed the resulting distributions will be. This suggests that a value of λ, even one scrupulously estimated from extensive and long-term data, will typically overestimate the *likely* future growth of a population. The use of λ to estimate the status of a population is acutely problematic in conservation biology because it will lead to overly optimistic prognoses of future population sizes. For these reasons, asymptotic and deterministic projections have very little real utility in conservation biology.

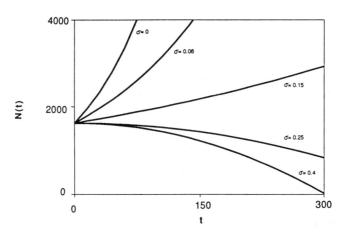

Figure 4.3 Population trajectories as a function of the standard deviation σ of juvenile survival rate p_0. The curves were fitted by non-linear regression to 30 replicates generated by Monte Carlo simulation. The initial abundance distribution used was 1000, 500 and 125 individuals in the zeroth, middle and last age classes. (Redrawn from Boyce, 1977.)

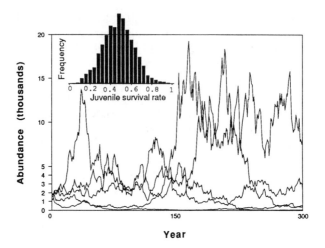

Figure 4.4 Five realizations selected at random from the Boyce simulation (Figure 4.3) using σ = 0.15. The inset shows the frequency distribution of juvenile survival rates observed during the simulation. A different value of p_0 was chosen from this distribution (at each of 300 time steps) to produce each of the trajectories. Only five realizations are shown so that the extraordinary dissimilarity among trajectiories is visible. Plotting more replicates yields wildly tangled trajectories and even more extreme outliers.

4.2.5 Reproductive value

Fisher's (1930) concept of reproductive value has a very natural expression in matrix models. Reproductive value is just the 'left' eigenvector **v** of the projection matrix corresponding to the dominant eigenvalue, defined by the equation

$$\lambda \mathbf{v} = \mathbf{v}\mathbf{L}. \tag{4.9}$$

(In contrast, the stable distribution is sometimes called the right eigenvector.) Reproductive value measures the worth of an individual in each of the age classes by the total number of progeny it can be expected to produce. This includes not only its immediate offspring, but all the future descendants of the individual. Since this total reproductive output can be infinite, the reproductive value schedule is usually expressed relative to the reproductive value of a newborn individual. Thus, reproductive value expresses the ultimate contribution to the population of an individual of a given class in newborn-equivalents. The total reproductive value of a population is defined as

$$\mathbf{v}\mathbf{n}(t) = \sum_{x}^{\omega} v_x n_x(t) \tag{4.10}$$

which amounts to the sum of the age-specific reproductive values weighted by the abundances in the various age classes. Unlike the total population size, which grows geometrically only at stable distribution, the total reproductive value for a population has the peculiar property of increasing or decreasing geometrically with every time step, no matter what the age distribution of abundance looks like.

Reproductive value was formulated by Fisher (1930) directly in terms of the demographic schedules and the population growth rate

$$v_x = \frac{\lambda^{x-1}}{\lambda_x} \sum_{y=x}^{\omega} \lambda^{-y} l_y m_y. \tag{4.11}$$

Occasionally this expression is still presented as the definition and formula for evaluating reproductive value. Besides being a bit cumbersome, the expression is tricky because its exact form depends on the definitions used in the model of population growth, which we argue below should reflect the life history of the species and details of the census. Since the formula includes λ, it assumes that the matrix has already been eigenanalysed. We think that defining reproductive value as the left eigenvector is somewhat simpler than Fisher's original formula, at least in this age of the computer (which will be doing the evaluation in any case). This definition also immediately generalizes for projection models more complex than the Leslie matrix (see Sections 4.3 and 4.6). Although it is neither foolproof nor very efficient, the following algorithm gives a fairly general way to estimate the schedule of reproductive values.

Algorithm 4.3 *Computing reproductive value*

1. Let Γ be the transpose of L.
2. Apply Algorithm 4.1 for the stable distribution to Γ.
3. The row vector given by transposing the estimate of the stable distribution is the reproductive value schedule.

Tailoring populations

When planning an introduction to re-establish a species in a location where it has become extinct, how can we judge whether it is better to bring juveniles or mature individuals? When culling individuals from a population that is in danger of overrunning its habitat, which age classes should be removed? Goodman (1980) argued that calculations with reproductive value could be used to make these decisions in an optimal way. For instance, it may not make sense to introduce or transplant individuals that have low reproductive value. Even if the operation is more expensive per transplanted individual, it may be worthwhile, and cheaper in the long run, to use individuals with higher reproductive value if the differential is large enough.

Age distributions that are far from the stable distribution will often undergo oscillations as they approach their stable structures. If the population cycling is strong, there may be times when nearly all the individuals are adults and times when they are nearly all juveniles. This may result in conservation problems. For instance, a population of all adults may put a greater strain on the environment and overutilize limited resources than a balanced population would. Goodman (1980) suggested that it may be possible to calculate the ideal stable distribution, especially in cases of highly managed populations where the necessary data are likely to be available. Then the actual distribution could be tailored by conservation and management practices so that its demographic cycling would be small enough not to damage the habitat in temporary overabundance or the population's viability in temporary underabundance.

The risk of population cycling may motivate us to reconsider our conclusions about how to stock wild populations to ensure healthy population growth. For example, a decision to transplant only individuals in the age group with the highest reproductive value would create a highly unbalanced population, far from its stable distribution. If the adverse consequences of the resulting population cycling would be too severe, we might choose to give up some reproductive value for an increase in stability of the population.

4.2.6   Deriving the Leslie matrix

In his original paper, Leslie (1945) proposed the projection matrix in the form

$$\begin{bmatrix} f_0 & f_1 & \cdots & f_{\omega-1} & f_\omega \\ p_0 & 0 & \cdots & 0 & 0 \\ 0 & p_1 & \cdots & 0 & 0 \\ \cdot & \cdot & & \cdot & \cdot \\ \cdot & \cdot & & \cdot & \cdot \\ \cdot & \cdot & & \cdot & \cdot \\ 0 & 0 & \cdots & p_{\omega-1} & 0 \end{bmatrix}$$

where f_x is the average number of daughters born between t and $t + 1$ per female in the x age class that survive to their first census at $t + 1$, and p_x is the fraction of those individuals of age x that will survive to be of age $x + 1$ at the next census. At the time of Leslie's work, most life histories that had been studied quantitatively were described in terms of the l_x and m_x schedules. For this reason, it was necessary for Leslie to show how to approximate the discrete schedules f_x and p_x from the information contained in l_x and m_x.

How the approximation should be done depends on just when reproduction occurs and how the census is made. For example, suppose the population is censused immediately before the birthing season, and that age is reckoned by the definition 'an individual is of age x if it has been alive between $(x - 0.5)$ and $(x + 0.5)$ years'. In this case, the expression for population structure next year is

$$n(t + 1) = \begin{bmatrix} p_0\left(\sum_{x=1}^{\omega} m_x n_x(t)\right) \\ p_1 n_1(t) \\ p_2 n_2(t) \\ \cdot \\ \cdot \\ \cdot \\ p_{\omega-1} n_{\omega-1}(t) \end{bmatrix} \qquad (4.12)$$

which means the Leslie matrix will have the form

$$\begin{bmatrix} p_0 m_1 & p_0 m_2 & \cdots & p_0 m_{\omega-1} & p_0 m_\omega \\ p_1 & 0 & \cdots & 0 & 0 \\ 0 & p_2 & \cdots & 0 & 0 \\ \cdot & \cdot & & \cdot & \cdot \\ \cdot & \cdot & & \cdot & \cdot \\ \cdot & \cdot & & \cdot & \cdot \\ 0 & 0 & \cdots & p_{\omega-1} & 0 \end{bmatrix}.$$

Compare this matrix to the expression we first derived in Section 4.2 (Equation (4.6)). Notice that the top row of this matrix has p_0 in each column instead of different p_x's. On the subdiagonal, the survival rates are shifted by one row.

Because different workers study species with different life histories, they formulate the approximations for the Leslie matrix elements somewhat differently. This has led to a good deal of confusion about what the elements of the Leslie matrix really are. Jenkins (1988) pointed out mistakes that seem to be made consistently by numerous authors. The problems arise in formulating the elements of the Leslie matrix from the l_x and m_x schedules. The first row of the Leslie matrix is not simply the vector of maternity rates m_x. These numbers characterize how many offspring are produced by an individual in each age class. Depending on the life history of the species and when the census is scheduled during the time step, two sources of mortality must be included in the Leslie model. Between the census and the breeding season, some number of the would-be mothers die. Also, between the breeding season and the following census, some of the offspring perish before they are counted. Both the pre-reproduction adult mortality and pre-census juvenile mortality must be accounted for in every properly formulated Leslie matrix. By scheduling the census either immediately before or immediately after the breeding season, one can assume either that pre-reproduction adult mortality is negligible, or that pre-census juvenile mortality is negligible. These census schedules lead to the common formulations seen for the Leslie matrix. If the census was not or cannot be scheduled for this convenience, it is essential to somehow account for both sources of mortality. Failure to do so can lead to overestimation of the growth of the population, yielding a falsely optimistic projection.

Our position is that there need not be a set dictum about what the elements of a Leslie matrix must be. We define a Leslie matrix to be a square matrix L that permits the projection of an age structure $n(t)$ (in which the age classes are one time unit wide) to $n(t + 1)$ by the equation

$$n(t + 1) = Ln(t). \tag{4.13}$$

Now, in general terms, the x, yth element of L measures the contribution of the abundance in the yth age class to the abundance in the xth age class in the next time step. But the best approximations for the various elements of L depend on

1. how age is determined
2. what the oldest age class is, and
3. when the census is made.

One constructs a Leslie matrix from a formulation that describes the life history of the species under study and the details of the census protocol. In practice, the Leslie matrix is determined by reference to an expression for $n(t + 1)$. How one estimates the elements depends on how mortality and reproduction are distributed throughout the seasons, and on what quantitative life history information is already available.

4.3 STAGE STRUCTURE

Leslie matrices are not appropriate for modelling all biological populations. Many species show a plasticity of development that makes age a poor indicator of demographic properties. A Leslie matrix will be inappropriate when the chances of reproducing, dispersing or dying depend on physiological development and an organism's status is independent of its age. This will be the case, for instance, when demographic rates depend on an individual's size and growth is controlled by environmental conditions. For many species, it is impossible to obtain a reliable estimate of age. If may only be possible to estimate some category or stage of development, be it based on age (young, middle-aged, old), size or some other character of the organism.

As an example, consider the life history of forest trees. Trees can spend decades in the understorey before an opening in the canopy allows them to grow to full size and become reproductive. Such events happen at different times around the forest, so the chronological age of a tree may end up having very little to do with its reproductive success. Virtually all trees die as seeds or seedlings and, because of overtopping, the size of an individual can be very important in determining success in intraspecific and interspecific competition. Thus, in trees, size is often the prime factor in both the reproductive and survival characteristics of the organism. It is cumbersome or even impossible to capture the population dynamics of such species in a Leslie matrix because doing so requires even-width age classes that have a simple relation with demography.

In this section we explore how we can relax the requirement that classes be age groups of identical duration. In the name of economy, we can re-use many of the same symbols we used for the age-structured models. Let

$$\mathbf{n}(t) = \begin{bmatrix} n_0(t) \\ n_1(t) \\ n_2(t) \\ . \\ . \\ . \\ n_\omega(t) \end{bmatrix} \tag{4.14}$$

stand for the stage-structured abundance distribution where the subscripts now indicate stage rather than age group. In principle, an individual in a stage can be of any age. For instance, the zeroth stage might be seeds, while the first and second are, say, seedlings and vegetative suckers. Other stages might represent taller saplings, understorey and canopy trees. We now seek an expression for the abundance $n_x(t + 1)$ at the next time step, and a matrix that will do this projection.

4.3.1 The Usher matrix

Usher (1966, 1969) introduced a matrix he constructed especially for studying the life histories of trees. It has the form

$$\begin{bmatrix} p_0 & f_1 & f_2 & \cdots & f_{\omega-1} & f_\omega \\ g_0 & p_1 & 0 & \cdots & 0 & 0 \\ 0 & g_1 & p_2 & \cdots & 0 & 0 \\ 0 & 0 & g_2 & \cdots & 0 & 0 \\ . & . & . & & . & . \\ . & . & . & & . & . \\ . & . & . & & . & . \\ 0 & 0 & 0 & \cdots & g_{\omega-1} & p_\omega \end{bmatrix}$$

where the diagonal elements p_x characterize the proportion of individuals in each stage who remain in that stage from one time step to the next. The subdiagonal elements g_x are the proportion that graduate to the next stage per time step. Caswell (1989) calls this matrix the 'standard size-classified matrix'. It is suitable for modelling many species in addition to trees, as the following example illustrates.

Loggerhead sea turtle
Caretta caretta is a long-lived iteroparous marine turtle. Like many marine turtles, it is a threatened species. To focus conservation efforts in the most efficient possible way, it is necessary to understand the demographic processes that determine its abundance. The turtles are very difficult to age because of several factors including their fast juvenile growth and their brittle shell that cannot hold marking tags.

Crouse *et al.* (1987) constructed a stage-structured model for the species using published data. They partitioned the population into

1. first-year individuals (eggs and hatchlings)
2. small juveniles
3. large juveniles
4. subadults
5. novice breeders
6. first-year remigrants and
7. mature breeders.

They pointed out in passing that the resulting 7 × 7 stage matrix makes computation considerably easier than was possible with a 54 × 54 age-classified matrix model that had previously been suggested in the literature. To parametrize a model with a yearly time step, they needed to estimate the annual reproductive output, the yearly probabilities of surviving and growing into the next stage, and the probabilities of surviving and remaining in the same stage. To estimate these numbers, they assumed that the population was stationary (despite suggestions it is declining) and that the age distribution of individuals within stages was stable. The resulting projection matrix was

	1	2	3	4	5	6	7
1	0	0	0	0	127	4	80
2	0.6747	0.7370	0	0	0	0	0
3	0	0.0486	0.6610	0	0	0	0
4	0	0	0.0147	0.6907	0	0	0
6	0	0	0	0.0518	0	0	0
6	0	0	0	0	0.8091	0	0
7	0	0	0	0	0	0.8091	0.8089

Notice that this matrix is similar to the standard Leslie matrix. Indeed the only difference is that some of the diagonal elements have non-zero values. These diagonal elements specify the proportion of the individuals in a stage class this year that will still be there in the following year.

Crouse and her co-workers used their stage-structured model to study the relative value of conservation measures designed to increase loggerhead abundance. Although most conservation efforts for the loggerhead sea turtle had focused on protecting the eggs on the nesting beaches, their results suggested the key to improving the demographic future for the loggerhead is to reduce mortality on juveniles, especially mortality on the large juveniles who are about to become reproductive.

4.3.2 The Lefkovitch matrix

In holometabolous insect populations, individuals can be grouped into eggs, larvae, pupae and adults. The demographic properties of an insect are largely determined by its developmental status which, in turn, is controlled by nutrition or temperature the organism has experienced rather than by its size or age. Insects pose another, purely technical, problem for an age-structured model. It is difficult to determine the ages of individuals, therefore it can be impractical to collect the data necessary to parametrize a Leslie matrix model.

Lefkovitch (1965) invented stage-structured models to avoid these practical problems in his modelling of insect population dynamics. The primary difficulty of the Leslie approach, as Lefkovitch saw it, is that the classification of individuals in the population has to be by age. Perhaps it would be better if, instead, we could adopt the natural divisions that individuals of a species fall into. For instance, we might keep track just of the abundances of eggs, larvae, pupae and adults rather than worrying about what actual age each individual had attained. In contrast to the Leslie matrix approach, in stage-structured modelling one can pick the groupings so they are composed of individuals who share demographically important characteristics.

In general, any sort of variable can be used to classify individuals, including developmental or morphological stage, size, social status, age, or perhaps some combination of these or other traits. Once we recognize the ecological or physiological traits that govern the demographic fates of organisms, we can construct a classification that will facilitate an efficient and accurate model of the population's dynamics.

Having decided how organisms should best be classified into groups, we soon find that the stages do not always fall into a simple sequence. We also discover that some species have complex transitions among stages that we did not encounter under the restrictive Leslie matrix. For instance, many plants exhibit vegetative reproduction in which an offspring directly enters a stage that is quite different from that of a seedling because it enjoys a much better chance of survival. There are even transitions in which organisms regress to an earlier stage when, for example, some environmental stress shuts off reproduction. We find that, in terms of stages, there is no longer a life cycle but more a network of possible interconnections.

Lefkovitch (1965) suggested that, in principle, the abundance of a stage could be influenced by any stage. Thus the expression for the next time step would be

$$\mathbf{n}(t+1) = \begin{bmatrix} p_{00} & p_{01} & p_{02} & \cdots & p_{0\omega-1} & p_{0\omega} \\ p_{10} & p_{11} & p_{12} & \cdots & p_{1\omega-1} & p_{1\omega} \\ p_{20} & p_{21} & p_{22} & \cdots & p_{2\omega-1} & p_{2\omega} \\ \cdot & \cdot & \cdot & & \cdot & \cdot \\ \cdot & \cdot & \cdot & & \cdot & \cdot \\ \cdot & \cdot & \cdot & & \cdot & \cdot \\ p_{\omega 0} & p_{\omega 1} & p_{\omega 2} & \cdots & p_{\omega\omega-1} & p_{\omega\omega} \end{bmatrix} \begin{bmatrix} n_0(t) \\ n_1(t) \\ n_2(t) \\ \cdot \\ \cdot \\ \cdot \\ n_\omega(t) \end{bmatrix}.$$

where the elements p_{xy} of the projection matrix are 'transition coefficients' that describe how the abundance of one stage will affect that of another. An element of a Lefkovitch projection matrix can represent a demographic event in which some number of the individuals go from one stage to another. In the simplest case, a transition can be the maturation of individuals in a stage to another class, such as seedlings growing into saplings. An element on the diagonal designates how much of the current abundance in a stage will remain in the same stage in the next time step. An element could also represent transitions in a more general sense. For instance, reproduction events are transitions from reproductive stages to newborn stages.

Even more abstractly, the element might describe the phenomeno-logical or statistical effect that abundance in one stage has on the abundance in another stage in the next time step. For instance, it might be a regression coefficient from a statistical analysis revealing a relation-ship of two stages between which no material actually flows. The p_{xy} are usually supposed to take only positive or zero values, since negative ones could lead to negative abundances.

A Lefkovitch matrix is a structural generalization of Leslie and Usher matrices in that any of the entries in the matrix may take on non-zero values instead of only those in the top row, diagonal and subdiagonal. Although, like Leslie matrices, the Lefkovitch model assumes a constant-width time step, the duration of time that an individual spends in each class is not necessarily the same as the time step. In fact the average time an individual spends in any particular class can vary from class to class. This allows a model to represent both seeds, for which average the residence time of individuals is one or two years, and canopy trees, for which residence times can be more than a century.

4.3.3   Criteria for stage-based modelling

The Lefkovitch model has since been used as the solution to a variety of problems in modelling population dynamics. For instance, a stage-structured matrix approach is often preferred over the traditional age-based Leslie matrix when

1. demography depends on physiological stage and development into these stages is not consistent in time among individuals;
2. demography depends on size or vigour and growth is plastic;
3. some individuals exhibit retarded or accelerated development or regression;
4. there are subclasses of the population that have different demographic characteristics (such as genders or spatially separated subpopulations);
5. the ages of individuals cannot be determined; and
6. equal-width classes lead to computational inefficiency.

The principal drawback of stage-structured modelling is the potentially large number of parameters that must be estimated. A few authors have compared age-based and stage-based approaches to studying and predicting population dynamics. Werner and Caswell (1977), for instance, concluded that a stage-structured model was clearly more accurate at predicting certain features of the demography of teasel plants. They suggest that this is likely to be the case for most sessile organisms which are selected to have plastic growth in response to environmental conditions. Despite initial pessimism about an approach based on body mass rather than age of ground squirrels, Sauer and Slade (1987) found the stage-structured model to be equal to or better than the traditional age-based model in capturing the important demographic characteristics. They also suggested that stage-structured modelling provided insights into certain dynamical properties of the population that were inaccessible with a Leslie matrix. Slobodkin (1953) and, more recently, Hughes and Connell (1987) suggested that both age and stage be included in a model; the way to do this is discussed in Section 4.6.

4.3.4   Asymptotic statistics

Several of the important statistics associated with the Leslie matrix can be computed for Lefkovitch matrices. In particular, the dominant eigenvector λ of a general stage-structured matrix can be interpreted as the asymptotic growth rate for a population growing according to that (unchanging) matrix. From Lotka's (1924) theory of population growth, we would expect the dominant eigenvalue from a Leslie matrix formulation and a Lefkovitch matrix formulation of the same population's dynamics to be equal. In practice, though, measurement error and disparities resulting from discretization can lead to different values. Furthermore, if both age and stage influence vital rates and the groupings are not in stable distribution, different estimates may also arise. It is even possible that estimates of λ by the two models can disagree about whether the population is increasing or decreasing (e.g. Werner and Caswell's, 1977 calculations for teasel, 'field L').

The eigenvectors for the stage-structured matrix also retain their interpretations. The left eigenvector measures reproductive value. In this case, of course, the values are expressed for each of the various stages, rather than age classes. The right eigenvector represents the stable stage distribution of abundances that would eventually be reached if the matrix were fixed through time. The conditions that guarantee that a population will asymptotically approach the stable distribution are somewhat more complex than for a Leslie matrix (Csetenyi and Logofet, 1989).

The stable distribution in capuchin monkeys

We used the results of a study by Robinson (1988) on groups of wedge-capped capuchin monkeys above as an example of the variability one might meet when trying to estimate life history parameters. He compiled information on the age, sex and reproductive status of individually recognized animals and used it to classify individuals into life history stages. The classes included juvenile and adult females, and juvenile, subadult and adult males. He used age- and sex-specific schedules of mortality and fecundity to estimate the stable distribution. He found that the distribution derived from the life table successfully predicted the observed average distribution of sex and age categories within the social groups. Robinson concluded that the group structure of the species could be explained by demography, rather than as a consequence of intragroup social interactions.

4.3.5 Sensitivity

Caswell (1978) introduced population ecology to an elegant tool for studying how a population's asymptotic growth rate depends on the elements of the projection matrix. Assuming the underlying projection matrix is fixed over time, he derived an expression for the partial derivative of the dominant eigenvalue λ with respect to the value of an element

$$s_{xy} = \frac{\delta\lambda}{\delta p_{xy}} = \frac{v_x u_y}{\sum_z u_z v_z} \tag{4.15}$$

where p_{xy} is the element in the xth row and yth column of the matrix and v and u are the left and right eigenvectors of the matrix, respectively. The value, called *sensitivity*, measures the impact that a small change in the matrix element would have on the asymptotic behaviour of the population. A large sensitivity value means that the corresponding transition element is important to the dynamics in the sense that modest changes in it would yield relatively large changes in the overall growth rate of the population. Sensitivities can be computed equally well for a Leslie matrix model or a more general Lefkovitch formulation. Of course, they can be especially useful for stage-structured models since such models can have numerous and different kinds of transitions.

Sensitivities could be very useful in planning empirical research. We would, for instance, be especially interested in getting a highly reliable measurement for a parameter whose sensitivity is high since any error in it would result in a comparatively greater inaccuracy in the model overall.

Sensitivity

Studying how a population's growth rate depends on variation in the vital rates has several important uses. First, it helps empiricists decide which vital rates to measure especially carefully. If a measurement error in a particular rate produces a large error in estimated growth rate, it is obviously advantageous to minimize the error in the measurement. Another important use is in planning optimal conservation or management strategies (Goodman, 1980). If there is a fixed level of investment that can be made in conserving a population, a sensitivity study could allow one to determine how this investment would best be distributed among alternative management schemes. For instance, is it better to increase ranger patrols to reduce poaching of adults, or would the extra costs of hiring rangers be better spent on habitat modifications that enhance survival of juveniles? Sensitivity studies can show which demographic phenomena are most responsive to fostering or protection from impact in terms of the overall population growth.

Goodman (1980) suggested that s_{xy} could be used in measuring sensitivity of growth rate to vital rates. While these values are often helpful to know, in practice, these analytic measures are not always adequate for this purpose. The values measure how the asymptotic growth rate changes under infinitesimal variation in the vital rates. Crouse *et al.* (1987) wanted to study the effects of 50% changes in the vital rates, which are rather too large to be considered infinitesimal. To do this, they just recomputed λ for each of a series of hypothetical projection matrices that had modified vital rates. They modified each parameter separately, and determined the modification that yielded the largest growth rate. Because λ is a complex function of all the elements of the matrix, the effect of changing two of the vital rates at the same time may not be estimable by sensitivities computed from the modifications separately. Thus, in general it may be important to compare joint modifications of two or more parameters if such management schemes are feasible.

It is important to remember that the value of a sensitivity study of population growth depends on the worth of the characterization used to measure population growth. We have described many of the limitations of λ as a conservation statistic. It is possible, of course, to use another benchmark instead of λ. Extinction risk, time to extinction, or chance of recovery are obvious choices that may be better suited for use in conservation studies.

Some of the sensitivities computed for a projection matrix can be very large numbers even though the corresponding entries in the matrix are zeros. This makes sense; if babies could have babies, the birth rate would be a great deal larger. But such a prediction does not seem very relevant. Some elements of the matrix may not be subject to change. There may be physical or biological constraints that keep the species from exploiting the increased growth it could effect by modifying these entries. De Kroon *et al.* (1986) suggested that proportional sensitivities or 'elasticities' be used to quantify how much the population growth is determined by a particular matrix element. The elasticity of an element of the projection matrix can be evaluated as

$$e_{xy} = \frac{p_{xy}}{\lambda} s_{xy}. \tag{4.16}$$

Elasticities are measures of the contribution each transition coefficient makes to the dominant eigenvalue. This interpretation is supported by the fact that all the elasticities in a matrix sum to 1. Since the column-sums are equal to the corresponding row-sums, they measure the total contribution of the class as a whole to the population growth rate, synthesizing its fecundity, survival and transition rates into a single scalar.

4.4 SIMULATING VARIABILITY

We have seen that it is often crucial to model the temporal variation in vital rates to get reasonable estimates about the future of population growth or decline. But exactly how should the fluctuation in vital rates be simulated in structured models? There are two fundamentally distinct ways: the *matrix selection method* and the *element selection method*.

Good years and bad years for jack-in-the-pulpit
Bierzychudek (1982) studied the forest perennial jack-in-the-pulpit at two sites in upstate New York. She partitioned the populations into stages representing seeds (I) and six size classes (II–VII) based

on projected leaf area. Reproduction in this species can occur sexually to produce seeds or asexually to produce stage II individuals. Through careful tagging studies, she followed individuals for three years and thereby estimated elements for two successive transition matrices.

Knowing that the vital rates for jack-in-the-pulpit change over time, she was not satisfied with calculating λ for each of the matrices, and pooling them did not seem to make sense. Following a suggestion by Cohen (1979), Bierzychudek used the observed matrices to simulate the possible future behaviour of the population by assuming the matrices represented demographically good and bad years for the population, which occur randomly but with equal frequency. Her Monte Carlo procedure simulated several statistical replicates of the current population. For each such population, she projected its future behaviour by selecting one of the two observed matrices at every time step. Thus each population might experience either of the two demographic conditions at any point in time. We refer to this approach as the matrix selection method for simulating the variation in vital rates. The outcome of this kind of simulation is shown in Figure 4.5.

An obvious alternative to the matrix selection method is to choose each of the vital rates independently, rather than as fixed combinations in a matrix. In the case of jack-in-the-pulpit, the resulting patterns of population decline are fairly similar whether one randomly selects yearly projection matrices from the set of matrices observed, or configures a new matrix each year by randomly sampling the vital rates observed. However this similarity need not always occur. Which method should be used in a simulation will depend on how environmental fluctuation induces variation in demography in a given situation. If there really are good and bad years that seem to control demographic success as a whole, then Bierzychudek's approach is reasonable. If environmental variation from year to year is more complex than this or if it triggers change in different vital rates independently, then the element selection approach to the simulation of the population's future may be warranted.

There can be systematic differences between the matrix and element selection methods. In the short term, the element selection method can yield a wider range for possible population size than the matrix selection method. This is because there can be recombinations of the vital rates that yield trajectories more extreme than any particular matrix generates. However, in the long term, these recombinations tend to cancel one

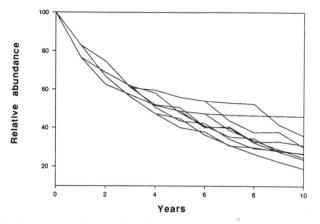

Figure 4.5 Ten population trajectories simulating the future abundance of jack-in-the-pulpit near Brooktondale, New York, over a ten-year period. Each realization started with the observed stage distribution in 1977. Thereafter, abundance was determined by a projection matrix selected at random at every time step from one of the two matrices observed over the 1977–78 and 1978–79 seasons (Bierzychudek, 1982, Table 3).

another out so that the distribution of population size after a long simulation with the element selection method is more concentrated than the distribution of population sizes produced using the matrix selection method. In short-term projections, the matrix selection method is constrained to a defined network of trajectories. Consequently there are gaps in the state space that no trajectory explores. This can be observed, for instance, in the first few years of the simulation for jack-in-the-pulpit. These gaps may or may not be plausible in any given circumstance. In the long term, the combinatorial possibilities from choosing a matrix at every time step begin to fill in these gaps. Such gaps rarely occur (or at any rate are relatively smaller) in simulations using the element selection method.

The objects that are randomly selected during the simulation, whether they are single elements or entire matrices, can be generated from reshuffling observed data collected from actual populations, or they can be synthesized based on parametric descriptions of the demography of the population. For instance, if the adult survival rates observed over a ten-year period are

$$0.78 \quad 0.76 \quad 0.53 \quad 0.40 \quad 0.76$$
$$0.32 \quad 0.73 \quad 0.58 \quad 0.66 \quad 0.80$$

we could use a random number between 1 and 10 to pick one of these values every time we need another value. We could also use a random number algorithm to generate values from a distribution with a mean of 0.63 and a standard deviation of 0.16. In principle, we could also model the observed skewness or any other statistical feature of our data set. This approach makes it easy to test hypotheses about the consequences of changing one of the statistical parameters. Choosing variates in this way forces us to make assumptions about the form of the underlying statistical distribution. And whether we believe it is normal or lognormal or some other distribution will determine which algorithm we use and the character of the variates it produces. Reshuffling has the advantage of not having to assume anything about the data (except that they are representative). The disadvantage is that limited data may not reflect the diversity of possible values the parameter may actually take on. Most importantly, they may not include the tails of the underlying distribution that produce extreme vital rates. Such values are important in estimating the risks of extreme events such as extinction.

4.5 CORRELATION AND AUTOCORRELATION

When vital rates vary stochastically they generally do not do so as statistically independent quantities. In fact, we often detect correlations among vital rates. For example, there may be correlations among several age classes with respect to a life history parameter. When survival of three year olds, for instance, is higher than average, we might expect the survival of four year olds to be elevated as well. There may also be pronounced correlation between reproduction and survival considered in aggregate. Whether such a correlation is positive or negative depends on the life history of the species. It could be that a good year for reproduction is also a good year for survivorship. The trend may go the other way such that when reproduction is higher than normal, it is at the expense of survival. Consider dispersal, which is often related to survivorship. In some populations, dispersal is high when survival rates are high, resulting in a positive correlation between these two parameters. In other populations, high local mortality induces migratory behaviour, resulting in a negative correlation between dispersal and survival rate.

Correlation among vital rates can have effects on the abundance trajectories we simulate for a population, and consequently on the conservation statistics we estimate from them. Therefore, it is important to include correlation in population models whenever there is evidence that it exists. For instance, if there is positive correlation among the vital rates in a matrix model, an element sampling method that does not reflect this correlation is likely to underestimate the extinction risk. This is because the random matrices that are formed during the simulation will not adequately depict the diversity of the transition matrices that actually occur. Whenever one observes statistically significant correlations among the vital rates of a population, precise estimation of its conservation statistics will require that these interactions be included in the simulation.

It is possible to generate random variates with a specified correlation structure for use in simulating the variability in vital rates. Clearly, however, this requires even more data (or guesses) than a model without correlation structure to estimate the correlation coefficients that should be used in the model. Unfortunately, even when all cross-correlations are measured, the task of simulating the deviates can be rather messy when several variables are involved. The Appendix provides algorithms to generate pairs and sets of correlated random numbers. More background information and algorithms for correlated random numbers can be found in Knuth (1981, p. 551). Some software exists to do this automatically for matrix-based population models (e.g. Ferson and Akçakaya, 1990).

Our intuition suggests that it is possible to make *conservative* estimates for conservation statistics without bothering to measure and then simulate all the correlations among vital rates. Conservative estimates (for example, an upper bound on extinction risk and a lower bound on time to extinction) can be made by assuming perfect correlation among the elements in the transition matrix. Perfect correlation may be simulated by combining a single random number for each time step with the respective means and variances of the several vital rates to produce the matrix used at a given step in the simulation. Using this approach one will produce trajectories that are maximally divergent for a given pattern of variances in vital rates. From such trajectories, the estimated extinction risk will be higher, and the estimated time to extinction sooner, than the actual population is likely to experience. A side benefit of this approach is that the assumption of perfect correlation allows the entire sample transition matrix to be computed from a single pseudorandom number selected for each time step. As a result, the simulations are relatively fast, and yield conservative bounds for extinction risk and time to extinction. If more accurate estimates of these conservation statistics are needed, one can expend the extra effort required both to measure the cross-correlations among the vital rates and to produce properly structured transition matrices for the simulation. We shall return to the idea of the conservative estimation of risk in Section 4.7 below.

There is another type of correlation that comes into play in matrix models that is also important in scalar (unstructured) models. There may be positive or negative autocorrelation in a vital rate over time. That is, there may be temporal patterns in the fluctuations of a vital rate that are induced by environmental trends or habitat conditioning. For unstructured population models, the sequence in which the vital rates fluctuate through time makes a difference only to the short-term estimates of population abundances. For time scales much longer than the scale of the autocorrelation, unstructured models are insensitive to the existence of autocorrelation. The same may not be true of structured populations. With age or stage structure, autocorrelated vital rates may induce very long lasting effects on population abundance. Tuljapurkar (1990, p. 81) mentions the impact that autocorrelation can have on various measures of population growth in structured models.

4.6 MIGRATION AND DISPERSAL

After the local extinction of a population, re-introduction of the species depends on the dispersal or immigration of individuals from other populations. Thus it is evident that the extinction dynamics at a site are mediated at least in part by the likelihood of such migration. And yet migration into or out of a population is neglected in most demographic models. Traditional formulations of the Leslie and Lefkovitch matrices assume that migration and dispersal are small or non-existent. This may be reasonable in some situations, but the assumption is mostly just a matter of mathematical convenience. Since they can play an important role in population extinction and recovery, we outline some ways to account for them in conservation models.

In Leslie matrix models, individuals either age into the next class or they die. In Lefkovitch matrix models, the possibilities further include transitions to any of the several classes or maintenance within the same class. In real populations, of course, there is yet another possibility. Individuals may simply leave the population. It is not unreasonable to think of this emigration as another kind of death. Indeed, depending on how survivorship is measured in the field, it may be impossible to distinguish the attrition of population numbers due to mortality and emigration. When this is the case, we often *do* include emigration in Leslie and Lefkovitch models even if unintentionally so; it is represented in the form of discounted values for the transition rates.

This way of representing migration in the projection matrix assumes that the number of migrants is proportional to the total size in the age or stage class. When the number of migrants is not a constant or stationary fraction of the class, this representation cannot accurately reflect migration. While there are infinitely many ways that migration may fail to be proportional, there is an important case that is easy to represent. Suppose that the *number* of migrants, rather than the proportion, is constant or statistically stationary. For instance, consider a situation in which the number of immigrants into a population is independent of its current abundance. If a fairly constant number of immigrants enter the population in each class every season, we could write this as

$$n_{x+1}(t + 1) = p_x n_x(t) + a_x \qquad (4.17)$$

where age-specific immigration is denoted by a_x. This would be a reasonable model when the number of immigrants into a population is independent of current population size and structure. This situation would be expected, for instance, if immigrants are dispersed passively or if immigrants cannot sense the conditions in the new population.

Like p_x, the value of a_x can be a stochastic variable in which case it will take on a different value at each time step. It represents the aggregate additions to the abundance in each age class from migration. Of course, the values could be negative as well. Negative values would represent constant subtractions from the population, such as when emigration is independent of population density. Such situations may be hard to imagine for natural populations, but often arise in connection with poaching and harvesting.

If mortality occurs later in the season, after migration, then the newcomers will suffer the same seasonal mortality as other individuals in the age class, and, instead of the equation above, one may write

$$n_{x+1}(t + 1) = p_x(n_x(t) + a_x). \qquad (4.18)$$

This difference can be important when gathering data for parametrizing the model. In any case, just as fecundity must be discounted to give the correct number of offspring that can be expected by the time of the next census, the empirical and formal definitions of 'migrant' must be consistent to yield an accurate model.

There is a way to arrange things so that this model can be captured in a matrix. For instance, consider an enlarged Leslie matrix formed by adding another row on the bottom (composed of zeros) and a column on the right of the matrix (made up of the elements in a_x). The new lower right-hand corner element contains a 1. This matrix is used to premultiply an augmented population vector formed by appending a dummy variable that always has the value 1 to the bottom of the population vector. For a Leslie matrix, this would be

$$\begin{bmatrix} n_0(t + 1) \\ n_1(t + 1) \\ n_2(t + 1) \\ \cdot \\ \cdot \\ \cdot \\ n_\omega(t + 1) \\ 1 \end{bmatrix} = \begin{bmatrix} f_0 & f_1 & f_2 & \ldots & f_\omega & a_0 \\ p_0 & 0 & 0 & \ldots & 0 & a_1 \\ 0 & p_1 & 0 & \ldots & 0 & a_2 \\ \cdot & \cdot & \cdot & & \cdot & \cdot \\ \cdot & \cdot & \cdot & & \cdot & \cdot \\ \cdot & \cdot & \cdot & & \cdot & \cdot \\ 0 & 0 & 0 & \ldots & 0 & a_\omega \\ 0 & 0 & 0 & \ldots & 0 & 1 \end{bmatrix} \begin{bmatrix} n_0(t) \\ n_1(t) \\ n_2(t) \\ \cdot \\ \cdot \\ \cdot \\ n_\omega(t) \\ 1 \end{bmatrix}. \qquad (4.19)$$

If mortality occurs after migration, then the column of a_x would be replaced by $p_x a_x$. Expanding this matrix equation shows that it is just the same system as above with the addition of an extra, degenerate scalar equation that sets one to one at each time step. You can use this matrix in any software you may have that does matrix multiplication to implement structured population models with additive effects. All the various conservation statistics can be computed from the augmented matrix model in a straightforward way. The quasiextinction times and risks for instance are computed in much the same way. The only precaution that must be taken is to be sure not to unwittingly include the dummy variable in any of the population summaries. Of course you must also be careful not to perturb the 1 at the bottom of the augmented population vector when you simulate an impact to the abundances.

Should you be interested in analysing this augmented matrix for its asymptotic properties (Caswell, 1989), you should be aware that the matrix is manifestly reducible. Therefore the asymptotic behaviour of the system must be studied by looking at the irreducible diagonal blocks of the matrix separately. In the case above (Equation (4.19)), this is fairly simple. The 1 in the lower right-hand corner of the augmented matrix is one of the blocks; the original matrix is the other block. The dominant eigenvalue for the 1×1 block is just 1. This means, for instance, that the long-term behaviour for a model with positive entries for migration can never decline to zero. At worst, it will decline to the level at which the immigration keeps it stocked. When the dominant eigenvalue of the original Leslie matrix is greater than 1, then its dynamics overwhelm the influence of the additive effects in asymptotic time. This occurs because the Leslie matrix causes the abundances to grow exponentially while the additive effects remain unchanging in magnitude on an absolute scale. When this is the case, the stable age distribution is a function of the Leslie matrix and the additive effects vector

$$\mathbf{u} = (\mathbf{I} - \mathbf{L})^{-1}\mathbf{a}$$

where $\mathbf{I}$ is the identity matrix.

4.6.1 Multiregional populations

In the models above, the 'away' to which emigrants go and from which immigrants come is undifferentiated. We do not keep track of what happens to individuals who leave the population, nor do we know the history of individuals who have just entered the population. In computing the overall conservation statistics, it may be very important to somehow include the dynamics of the other populations that provide immigrants. We can be more detailed about the other populations with which migration occurs by constructing a complex stage-structured model that includes both the internal dynamics of each population and the migration among populations. Rogers (1966, 1968, 1985) introduced and elaborated these 'multiregional' models of growth for human populations.

Dispersal and extinction risk
Fahrig and Merriam (1985) introduced a multiregional model for the population dynamics of the white-footed mouse, *Peromyscus leucopus*, which lives in distinct patches that are connected by dispersal. Their population vector recorded the abundance in each of three age classes within four patches as

$$\left.\begin{array}{l}n_{1,1}\\n_{2,1}\\n_{3,1}\end{array}\right\}\text{patch 1}\qquad\left.\begin{array}{l}n_{1,2}\\n_{2,2}\\n_{3,2}\end{array}\right\}\text{patch 2}\qquad\left.\begin{array}{l}n_{1,3}\\n_{2,3}\\n_{3,3}\end{array}\right\}\text{patch 3}\qquad\left.\begin{array}{l}n_{1,4}\\n_{2,4}\\n_{3,4}\end{array}\right\}\text{patch 4}$$

where the first subscript denotes age and the second denotes patch. These four sets of numbers make a vector with 12 elements. Four copies of a 3×3 Lefkovitch matrix describing the dynamics within each of the patches formed the diagonal blocks of the 12×12 transition matrtix. The off-diagonal blocks described age-specific migration among the various patches. The vital parameters for the diagonal matrices were estimated from data reported in the literature and the migration pathways were estimated from observations on dispersal rates measured in the field. In *Peromyscus*, dispersal, as well as mortality and fecundity, are evidently density-independent.

The researchers assumed that some proportion of the animals disperse whether or not there is another population nearby for them to enter. This is reasonable since the animals probably have no way of sensing conditions far from their current habitat. Dispersing individuals who do not re-enter any population were assumed to be lost to predation. The model was stochastic so the values in the matrix could change from week to week. The model predicts that a multiregional population of the white-footed mouse is less prone to extinction than nearby, isolated populations that experience no cross-dispersal. The overall short-term growth of a population that is part of a network of cross-dispersing populations seems to be greater than for an isolated population (Figure 4.6).

This prediction is confirmed by empirical observation. It is attributable to the mortality of animals who dispersed from isolated populations who never re-entered a population. This suggests that, although dispersal can improve the chances of population survival when a collection of populations are connected together, when a population is isolated from other populations, dispersal acts to increase the risk of and shorten the time to extinction.

This section has discussed elaborations of the matrix approach that can be used to model proportional or additive effects of migration and dispersal. In nature, migration is rarely as simple as in either model. Sometimes migration is especially strong when population density is very high. In other species, migratory behaviour is triggered when density becomes very low. In general, when migration is determined by some non-linear function, the population model must be more complex. Section 4.7 discusses ways to introduce such density dependence into matrix models. Pollard (1966) explored other ways to model migration with stochastic matrix formulations. Chapter 5 on spatial models will introduce a variety of methods that treat the influences of migration and spatial

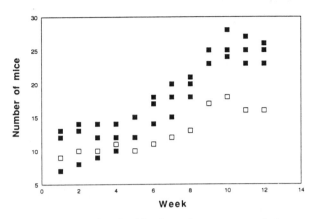

Figure 4.6 Predicted growth of white-footed mouse populations experiencing cross-dispersal with nearby populations (solid squares) or isolated from other populations (open squares). (After Farhig and Merriam, 1985.)

complexity in considerably more detail. It describes variously detailed ways to model the population dynamics of an entire system of populations that may exchange individuals. Such 'metapopulations' have dynamical properties that can result in conservation statistics radically different from those of a single population.

4.7   DENSITY DEPENDENCE

A fixed projection matrix allows only a few possible outcomes from deterministic population growth in asymptotic time. The population may experience a geometric decline to extinction, a geometric increase to infinity or an exactly balanced stationarity. The conditions giving rise to balance are delicate and may be destroyed by small perturbations. Permanent cycles may accompany an increase, decrease or balance, but permanent cycles are also delicate in that they can be destroyed by small changes in the elements of the projection matrix. Given that some age-structured populations have evidently neither grown to infinity nor fallen to extinction, some phenomenon other than those encapsulated in the traditional Leslie matrix formulation must account for this.

Dispersal, which recolonizes locally extinct populations from other, luckier populations, and environmental fluctuation, which varies the vital rates so that a population is sometimes increasing and sometimes decreasing, have both been suggested as possible mechanisms that keep populations sizes at intermediate levels, neither extinct nor infinite. Another possible mechanism is *density dependence*, by which populations are constrained to intermediate abundances by some kind of feedback of population size on the vital rates. As it can in the case of unstructured populations, density dependence in age- or stage-structured models can strongly influence a population's dynamics and the resulting conservation statistics we estimate for it. In this section, we shall review ways to build density dependence into matrix models, and discuss the prospects of doing so with the kind and quantity of data that are usually available.

Leslie (1948, 1959) first incorporated density dependence into a matrix model of population growth. He explored various ways of systematically decreasing the vital rates in the projection matrix so that the population tended to a fixed carrying capacity. His approach has not been widely used, however. Primarily, this has been because biological species never seem to cooperate in exhibiting orderly covarying vital rates. Recall that density dependence is a very broadly defined term. Whenever change in population density or abundance alters the nature of any demographic process, the phenomenon is called density dependence. This definition suggests a straightforward way to incorporate density dependence into a matrix model of population dynamics. We can replace any of the constants or stochastic variables in the projection matrix that specify the various demographic processes with functions that depend on the population size. As the population varies over time, such a function reflects how the transition coefficient varies with respect to the population's abundance. For example, if we know empirically that high population density reduces the survival rate for an age class, then the element in the projection matrix that describes that survival rate can be replaced with a function that yields a low value when the density is high. At every time step, before multiplying the abundance vector by the projection matrix, the function is evaluated based on the current population abundance. Pennycuick *et al.* (1968) suggested this approach and explored the dynamical consequences of using particular functions to describe this density feedback on vital rates. Pennycuick (1969) used the approach to study the regulation of an English population of the great tit *Parus major*.

Allee effects and extinction risks

If the biology of a species is such that when a population's abundance falls to a low level certain factors depressing population growth begin to act more strongly, the population is said to experience Allee effects. In some cases these factors may cause a small population to fall even lower in abundance, perhaps leading to an 'extinction vortex' (Gilpin and Soulé, 1986). Estimating extinction statistics for such a population with a model that does not include these Allee effects will be overly optimistic and therefore misrepresent the danger of extinction that the population faces (see Chapter 3).

Beddington (1974) offers an example of how to incorporate Allee effects into a projection matrix. In a model of the population growth of an insect species in a laboratory culture, he replaced fecundity and survival rates of a Leslie matrix by functions of total population size N. The result was

$$\begin{pmatrix} 0 & 0 & f(N) & f(N) \\ p(N) & 0 & 0 & 0 \\ 0 & p(N) & 0 & 0 \\ 0 & 0 & p(N) & 0 \end{pmatrix}$$

where

$$f(N) = a \ln N + b(\ln N)^2 + c$$
$$p(N) = d - e \ln N$$

and the parameters for these functions were fitted to empirical observations by least-squares regression. Beddington warned that these fitted functions produced meaningless results when N is outside a certain range, but he argued that they describe how reproduction and survivorship seem to vary with different population densities. Because the fecundity function is strongly non-linear, the future of a population trajectory depends whether it goes lower than the unstable point N^{**} (Figure 4.7). When above this point, the population tends to a stationary size around 600 individuals. If abundance starts below this critical level, or slips below it by chance fluctuations, it will tend to extinction.

The probability of a population's extinction obviously depends on whether such critical levels exist. Allee effects induce a kind of gravity that draws trajectories down to extinction. In conservation biology, the presence and importance of Allee effects in natural populations is often realized only after they have been expressed.

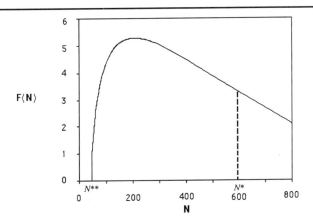

Figure 4.7 Density dependence in fecundity rates for an insect population. N^* marks an equilbrial population size; N^{**} marks the critical size below which depressed reproduction caused population extinction. (After Beddington, 1974.)

Because of its broad definition, density dependence may influence any of the coefficients characterizing reproduction, survival or dispersal. The approach of replacing matrix elements by functions is very flexible. Caswell (1989) suggested that any non-negative function may be used. There are certain constraints that should be kept in mind, however. For Leslie matrices, one need only remember that survival rates have to be less than or equal to 1. In general, for Lefkovitch matrices, this means that the sum of all non-reproductive transitions must be less than or equal to 1. It can be tricky to build these constraints into Lefkovitch models.

The flexibility of the tactic of replacing matrix elements with functions is, in fact, the most serious problem with this approach to modelling density dependence in matrix models. There are three decisions one faces on introducing non-linearity into the model. First, one has to choose the mathematical form of the function. This decision is obviously crucial in determining the dynamical behaviour of the resulting population trajectories. Unlike unstructured models, one also has to decide what the argument of the function is to be. In the models of Chapter 3, the argument was always the current population size. In structured models, however, one has to choose whether it is the total population, or the abundance in a particular class or set of classes that controls the feedback mechanism being modelled. Perhaps the abundances in the various age classes should be weighted by their respective body sizes. Choosing the argument intelligently requires a detailed life history of the species. Finally, one has to decide how to estimate the parameters of the density-dependence function(s) that have been incorporated into the model. Again, a structured model can be considerably less tractable if several parameters need to be estimated from limited data.

Self-shading in kelp

The giant kelp *Macrocystis pyrifera* grows in lush forests off the coast of California. These forests are ecologically important because they provide habitat for a variety of coastal invertebrates, fish and mammal species. Kelp is also an economically important species that is harvested for many human uses.

In the 1980s, a nuclear power plant was blamed for the local collapse of the kelp population at certain sites. To study the population dynamics of the giant kelp, Burgman and Gerard (1990) developed a stage-structured matrix model. Their formulation allowed the elements of the projection matrix to vary stochastically with environmental parameters such as water temperature (a surrogate for nutrient availability) and light incidence which were known to have a strong influence on growth and recruitment in the species. Temperature and solar irradiance were treated as driver variables, reflecting seasonal variation throughout the year. Burgman and

Gerard postulated that the recruitment and growth of each kelp plant depends, as they do in any forest, on the amount of light that actually reaches the plant.

Using data on kelp growth, they derived empirical functions relating the growth of understorey kelp to the amount of water that penetrates the water column through the overstorey plants. When they ran simulations of kelp population growth, they observed distinct temporal windows of massive recruitment, separated by periods where the gamete formation by kelp plants is strongly limited by shading from older plants. In older stands, this produces a lopsided population distribution of many large plants and few immature plants. With such a distribution, ocean storms can decimate or completely destroy a stand because adults are particularly susceptible to wave action. Their model suggested that significant reductions in local kelp abundance are natural phenomena induced by density dependence in the population growth of kelp.

The dynamics of the kelp forest are difficult to predict precisely because weather patterns cannot be accurately forecasted from year to year. But the correlation between low adult density and high juvenile recruitment has been widely documented for other kelp forests, and kelp forests evidently also experience population fluctuations at locations far from the nuclear facility.

4.7.1 Density vagueness

When researchers collect data to estimate the form of density dependence, it is extremely rare that the data points arrange themselves into clear patterns. In many cases, perhaps the most surprising thing is the degree of their scatter (e.g. Figure 4.8). In characterizing density dependence, and in building models to predict its consequences, the best minds in population biology have taken what can only be described as leaps of faith in drawing lines through shotgun blasts of data. While these efforts have been essential in the search for systematic explanations demanded by science, the regressions are of little use in the modelling needed by conservation biology. This is because the resulting curves and precise models ignore *variability* in density dependence.

To assess risks and to plan conservation strategies, we need to treat the variation itself as the important thing. It is probably silly to pretend there exists a perfect density-dependence curve in a scattergram of data and tortuously thread some guess at its form through a Leslie and Lefkovitch matrix. Often there is very little in the data that suggests the form of the equation the researcher should use for the regression in the first place. One strategy that is very useful in this case (and is also quite easy to

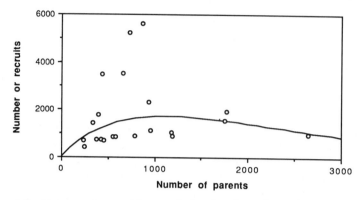

Figure 4.8 Data used as evidence of density dependence in Tillamook Bay salmon (*Oncorhynchus keta*), and the Ricker recruitment curve fitted to these points. Despite considerable fluctuation in the data points, curves such as the one shown are traditionally used to represent density dependence. (After Ricker, 1975.)

implement in simulation modelling) is to use the data themselves as the encapsulation of density dependence. For instance, suppose we have observations on the dependence of some demographic statistic on population size. For a given population size, there have been several values of the statistic that have been measured at various times. Rather than assuming that the average or predicted value is the best one to use when a simulated trajectory achieves that population size, instead one can select a value at random from those values actually observed. This procedure is analogous to selecting adult survival rates from a list of observed values, a method we suggested in Section 4.4 above to represent environmental variation. It projects the variability that we observe in the density dependence data directly into the estimation of conservation statistics.

4.7.2 Conservative estimations

Because density dependence is difficult to detect, and because it requires a great amount of data and understanding to determine the correct mathematical form and values for the parameters of density-dependence equations, it is tempting to ignore the problem. The question then arises: What happens if density dependence is omitted from a model? In the deterministic case, populations usually will either grow or decline exponentially, once the stable age distribution has been attained. The mean trajectory of a stochastic matrix model will do the same thing; however Ginzburg *et al.* (1990) found that the omission of density dependence can be a useful tool in evaluating quasiextinction risks.

The probability of extinction of a population is sensitive to the presence and form of density dependence. In the examples of the unstructured models for suburban shrews in Chapter 3, the Ricker and logistic equations for density dependence gave very different estimates of the chances of extinction of the population, even though the parameters for both were based on the same set of observations. Ginzburg *et al.* (1990) used data from a population of the cod *Gadus morhua* and found that the parameters of the Ricker function were difficult to estimate accurately, even for relatively long sequences of data up to 40 years. Furthermore, the predictions of an age-structured model were very sensitive to estimates of the strength of density dependence. Because of measurement error in estimating the parameters of the Ricker function, an estimate that is two standard deviations from the true value may translate into as much as a five-fold error in risk estimation.

Their simulations would suggest that quasiextinction is too unstable an index to be useful in monitoring and evaluating natural populations whenever data for reconstructing density dependence are limited, were it not for an interesting feature of the results. They found that the quasiextinction risks for the model without density dependence were quite high. In general, compared to models with plausible estimates of density dependence, the density-independent simulations gave *conservative* estimates of risk for most low thresholds. Because we are concerned with the risk of an undesirable event such as a population's reduction to extinction, it is useful to evaluate the upper bounds on the probability that it will occur. In the case of the Beverton–Holt equation or the Ricker equation with moderate non-linearity, the density-dependence mechanism is a restoring force that tends to return the population to an equilibrial size whenever it deviates because of environmental perturbations, thereby reducing the risks of quasiextinction.

The true quasiextinction risks faced by a population will be no greater than the risks estimated using models without density dependence, as long as relatively small threshold sizes are considered. This relationship may break down when density dependence is very strong because limit cycles and deterministic chaos generate very high quasiextinction risks in models of scramble competition such as the Ricker equation. There is no guarantee that density dependence will necessarily modify the elements of a matrix in such a way that the growth rate tends to stabilize (*contra* Emlen, 1984, p. 92). An important counter-example is when Allee effects destabilize population growth rates. But whenever density dependence is stabilizing, a density-independent model may be used to generate conservative estimates of quasiextinction.

4.8 CONCLUSIONS

In the previous chapters, we represented a population with a single number, *N*, which varies over time. In matrix models a population can have internal structure and we capture this fact with a series of numbers representing the sizes of the components of the population, however these components are defined. Matrix models can be very detailed and quite flexible. Their primary limitation is that, as matrices, they treat all processes in discrete time.

The asymptotic finite rate of population growth, measured by the dominant eigenvalue λ of the projection matrix, is the traditional focus of population studies using matrix models. Many authors admit that the concept has limitations but stress it is a synoptic and generally useful measure of the future of a population. Keyfitz (1972) made a valid distinction between *forecasting* what will happen and *projecting* what would happen if conditions were to remain fixed. This encouraged Caswell (1989) to champion a variety of asymptotic and often deterministic results as potent tools by which biologists can study a population. He suggested that these tools are projections that reveal essential features about the population because they integrate the impact of the environmental conditions on the life cycle and the associated vital rates.

While asymptotic and deterministic analyses may be an interesting summary of current demography, forecasting with λ is irrelevant in most applications of interest in conservation biology. Demographic studies in conservation biology are concerned with several kinds of computations, such as the risk of extinction or population decline, the likely time frame for these events, the chances for population recovery, the minimum population likely to survive the near term, and so forth. It is because of the unique kinds of problems in conservation biology that deterministic and asymptotic analyses have limited utility. In the short term, the initial population distribution can strongly influence the future population. This can render λ irrelevant if extinction or the complications of quasiextinction are likely in the short term. In the long term, the vital rates that determine λ are virtually certain to change under fluctuations in the natural environment. In any case, the noise from both demographic and environmental stochasticity can lead to extreme events that pre-empt all predictions about asymptotic behaviour.

The utility of λ is also complicated when it is generalized for a stochastic matrix model. Although an average population increase has the arithmetic mean of λ, the probability of seeing any particular population increase at a rate other than the geometric mean is asymptotically zero. It is even possible for the mean growth rate to suggest that a population will increase while nearly every population from the statistical ensemble of trajectories becomes extinct. In such a case, the 'growth rate' represented by λ is clearly unrepresentative of a population's future.

This result is analogous to that for the unstructured models explored in Chapters 1, 2 and 3. The distribution of N is skewed so that the most probable trajectory is below the average. Thus, relying on the expected growth rate as a measure of the robustness of a population will be overly optimistic. The farther into the future the prediction is made, the more overly optimistic it becomes. Direct simulation is a straightforward alternative to analytical asymptotic analysis. It usually requires a computer to make useful computations, but simulations are amenable to considerable elaboration. For instance, it is relatively easy to introduce non-linear density dependence into a stage-structured simulation model when such complexity would make analytical models of population growth intractable.

Caswell (1989) is the primary reference for modelling structured populations with matrices. Caswell gives thorough coverage to many essential topics, including asymptotic dynamics, eigenanalysis, analytic methods to study sensitivity, and transient dynamics. He also discusses ways to estimate the parameters needed for age- and stage-structured models. He does not discuss the problem of how to choose the categories of the structure (which turns out to be a very important decision). For the mathematically inclined, Tuljapurkar (1990) is a surprisingly readable discussion of the theory of structured populations growing in stochastic environments. Keyfitz (1968) treats continuous-time models in which the intervals of age are considered to be very small and notes that 'some results are not accessible to matrix theory'. Getz and Haight (1989) address questions arising from managing populations for harvest. The mathematics for the dynamics of exploited populations is quite different from that developed for descriptive biology or conservation biology.

The statistical packages Minitab (Ryan *et al.*, 1985) and SAS (SAS Institute, 1985) both support matrices and a variety of standard manipulations that make matrix calculations quite easy. In the microcomputer environment, working with matrices is probably most natural in the programming language True BASIC (Kemeny and Kurtz, 1985). The computation needed for population projection with the models described in this chapter becomes fairly easy with such software tools. RAMAS/age (Ferson and Akçakaya, 1990) and RAMAS/stage (Akçakaya and Ferson, 1990) are specialized programs that will compute quasiextinction and quasiexplosion risks from age- or stage-structured demographic models.

4.9 SUMMARY

Structured models account for differences among individuals in a population. Abundance and survival, fecundity and dispersal rates must be carefully defined and the structure of the model should reflect these definitions. Leslie matrices encapsulate survivorships and maternity rates in age-structured populations. The stable distribution and the asymptotic growth rate, λ, of an age-structured population can be determined from the Leslie matrix; they are the dominant right eigenvector and eigenvalue of the matrix, respectively. The reproductive value is the associated left eigenvector. The initial distribution may have a very important effect on the short-term behaviour of a population. Random variation in the matrix elements may be employed to represent the impact of stochastic environmental variation on the vital rates of a population.

The Usher matrix is a special case of the generalized Lefkovitch matrix for stage-structured populations. These matrices are described, and Caswell's (1978) equations to calculate the sensitivity and elasticity of transition matrix elements are provided. Like the Leslie matrix, random variation in matrix elements may be used to represent the effects of environmental variation on a population. Matrix and element selection methods are alternatives to generate conservation statistics for structured populations.

Correlation among the vital rates of various age or stage classes, and autocorrelation among vital rates through time, may have important effects on the estimation of the risk of quasiextinction and the time to extinction. Autocorrelation is a less important consideration in unstructured models. Conservative estimates of risk may be obtained by assuming perfect correlation among matrix elements.

There are a number of ways in which migration and dispersal may be accommodated in matrix population models. The method should be tailored to reflect the biology of the species. Matrix models are also sufficiently flexible to allow non-linear density dependence. Density dependence may be difficult to detect and to parametrize unless long sequences of reliable data are available and there is good understanding of the density-dependence mechanisms. Conservative estimates of conservation statistics may be obtained by ignoring density dependence if the density-dependence mechanism stabilizes population growth, and if relatively small threshold population sizes are considered.

Many authors have used the quantity λ to evaluate the health of a population and to predict its fate under various management scenarios. The use of λ in this role is intimately intertwined with the idea that it is meaningful to project the present abundance distribution infinitely far into the future by assuming that the projection matrix does not change over time. We have explored a variety of reasons why this is unreasonable, and have sought alternative approaches useful for conservation biology.

5 Spatial structure and metapopulation dynamics

The previous chapters have concentrated on models of single populations living in habitats that vary through time. In all of these models we assume that the environment is uniform in space. That is, we assume that all individuals, no matter where they occur within the range of a population, experience the same changes in the environment and have the same chances of surviving and reproducing.

Few biologists feel entirely at ease with this assumption. Quite apart from spatial variation in variables such as soil conditions and elevation, more extreme events including fires, droughts and floods affect established populations. Rarely are the changes wrought by these events uniform throughout the landscape, and how an individual fares will depend critically on where it happens to be. For example, fires burn in mosaics that depend on fuel loads, moisture conditions, natural barriers and prevailing winds. Different parts of the habitat burn at different intensities and with different frequencies, and some parts escape fire altogether.

This chapter deals with spatial structure, which adds another dimension to viability analysis. When spatial structure is included in a model, we account for the fact that individuals in the same population experience different living conditions. When we built models to account for density dependence, environmental and demographic variability, and age, and stage structure in the preceding chapters, we organized our ideas in a framework that allowed us to build models that incorporate the various phenomena. It is important to do the same for the spatial structure of populations.

Spatial heterogeneity in the environment may result in a patchy distribution of organisms. The result is that local populations occupy patches of high-quality habitat and use the intervening habitat only for movement from one patch to another. Many species exist in a number of populations that are either isolated from one another or have limited exchange of individuals. Such a collection of interacting populations of the same species is called a **metapopulation**. This concept underpins the framework for building models that incorporate spatial variation.

The effects of environmental and demographic stochasticity that we emphasized in previous chapters become more pronounced when populations decrease in size as a result of habitat loss and fragmentation, leading to frequent local extinctions and recolonizations. This may prevent each population from reaching its equilibrium and staying there, although the metapopulation as a whole may persist much longer than any individual population. This characteristic of natural populations, known as non-equilibrial dynamics, has long been recognized as an important factor by ecologists (see, for example, Andrewartha and Birch, 1954). The turnover of populations produces a temporal pattern referred to as a shifting mosaic, which is a pattern of occupancy of local populations in a metapopulation. This pattern of occupancy changes over time as patches become extinct and are recolonized.

The patterns of occupancy are made more complex by differences among populations in terms of their carrying capacities, growth rates, the magnitudes of environmental fluctuations they experience, and the rates of migration or dispersal among patches. We use these latter two terms interchangeably. In some cases, differences among populations may be very important for the overall dynamics of the metapopulation. For example, differences in the productivity of patches may lead to populations that receive migrants but seldom produce any offspring or send emigrants to other populations. These habitat patches act as sinks, absorbing emigrants from other patches.

Local extinctions

Furbish's lousewort (*Pedicularis furbishiae*) is an endangered plant that is endemic to northern Maine (USA) and adjacent New Brunswick (Canada). It was assumed to be extinct for 30 years until its rediscovery in 1976. It is now known to exist in 28 populations along a 140-mile stretch of the St John River. The dynamics of this metapopulation are characterized by frequent local extinctions caused by disturbances such as ice scour and bank slumping, which are distributed patchily (Menges, 1990). These disturbances also seem to be essential for the species' survival since they prevent tree and shrub establishment which depress lousewort populations. As a result, individual populations are short-lived, with fairly rapid increases followed by catastrophic losses. This natural disturbance pattern makes the viability of the species dependent on dispersal and establishment of new populations (Menges and Gawler, 1986; Menges, 1990).

Spatial heterogeneity and the dynamics of multiple populations are usually omitted from population models. One reason is that many population models are concerned with dynamics within confined areas, such as a single nature reserve or a lake. Another reason for the omission of spatial structure has been that it is quite difficult to model. Tractable and realistic models of spatial dynamics have been developed only recently. Spatial structure and metapopulation dynamics have important consequences for viability analysis and extinction risk assessment. There are four main reasons for this.

1. Conservation biology is often concerned with the protection of species, not just single populations. To determine how threatened a given species is, it is not enough to estimate the risk of extinction for only one (or even separately for all) of its populations. One cannot estimate the risk of extinction of an entire species based on the risk of extinction of its separate populations because of the interdependency of and

the interactions among the populations. For the purpose of species conservation, population viability analysis introduced in the previous chapters must be extended to viability analysis that incorporates these complex interactions.

2. Habitat loss is probably the major cause of species extinctions. When land is cleared for agriculture, for example, this often results in fragmentation of habitat, which creates a metapopulation from a single large population. Metapopulation dynamics are especially important for endangered species, many of which exist in such small, relatively isolated populations that have resulted from habitat loss and fragmentation.

3. The existence of multiple populations introduces new susceptibilities. In addition to impacts that decrease the mean survivorship or fecundity in a single population, a metapopulation can be subject to impacts that affect movement of organisms and increase the isolation of populations, such as road building, construction of dams, agriculture and so on. There may be other factors that decrease isolation, such as wildlife managers translocating individuals among populations and re-introducing populations to empty patches. These activities will introduce the need to make new kinds of decisions such as how to formulate the schedule of translocations, or whether to re-introduce a large number of individuals to a single patch or smaller numbers of individuals to several patches.

4. An important aspect of reserve design involves selecting patches of habitat for protection of a variety of species. Which combination of nature reserves gives an endangered species the highest chance of survival can only be assessed by an analysis of metapopulation dynamics. Similarly, the question of whether a single large reserve or several small reserves of the same total area provide better protection can be answered only on a case-by-case basis using species viability analysis.

Implicit in points 2 and 4 above is the assumption that conservation usually is reactive to habitat loss and fragmentation. Conservation can be proactive, dealing with the creation and restoration of habitat (Janzen, 1988; Jordan *et al.*, 1990). Considerations of spatial structure are crucial in the design of new habitat.

Factors that affect population extinction risks include population size and structure, life history parameters, demographic and environmental stochasticity that cause variation in these parameters, and the correlations among life history parameters within the population. These topics have been discussed in previous chapters in relation to single populations. Species extinction risks depend on all the factors that affect population extinction risks and, in addition, on other factors that characterize the interactions among populations. The additional factors that operate at the species level include the number and spatial configuration of populations in which the species lives, the similarity of the environmental conditions that the populations experience, and dispersal among populations that may lead to recolonization of locally extinct populations.

In this chapter, we shall review natural and human-induced phenomena that are important in determining species extinction risk at the meta-population level. We shall give examples of empirical metapopulation studies that have been reported in the literature. We shall then discuss the development of metapopulation models within two broad categories, namely occupancy models and population dynamic models. The latter approach is quite new, and we believe it provides the most realistic approach to evaluating the importance of spatial structure. We therefore discuss it in some detail and provide examples to illustrate the importance of interactions between geographic configuration of populations, dispersal patterns and covariation of environmental patterns.

Dispersal effects

Southern California populations of Mountain sheep (*Ovis cana-densis*) inhabit mountain 'islands' in a desert habitat. Bleich *et al.* (1990) give an example of such a metapopulation, in which about 1000 mountain sheep inhabit 15 of 31 mountain ranges with 6 to 20 km between any two ranges. Although the mountainous patches are separated by unsuitable habitat, Bleich *et al.* have documented movement of mountain sheep between 11 pairs of these mountain ranges. In addition, two of these patches have been re-established by translocation. The authors conclude that the movement of sheep among mountain patches is important for their conservation from both genetic and ecological points of view.

5.1 COMPONENTS OF SPATIAL STRUCTURE

Metapopulations occur because the environmental factors necessary for the survival of the species occur in patches. For example, giant kelp (*Macrocystis pyrifera*) in the coastal waters off southern California grow in patches determined mostly by the properties of the substrate on the ocean floor, exposure to wave action and water depth (Burgman and Gerard, 1990). The species-rich sand heaths of Western Australia are restricted to patches of suitable substrate, isolated by weathering of tertiary sediments (Hopper, 1979). Islands in an archipelago are another example of habitat distributed in spatially separate patches.

When a species has several populations, the extinction probability of the entire species will be a function of the number of populations. Consider a case of a species that consists of two populations far apart and completely isolated from each other. In this case we might assume that the extinction probabilities of the two local populations are independent. If each of these populations faces a 10% risk of extinction within a specified time, the extinction probability of the species as a whole will be 1% over the same period of time.

We can calculate this probability because when two events are independent, the joint probability that both will happen is the product of their two constituent probabilities. It is essentially the same as tossing two coins. Hence, if we consider only the number of populations, a species with several isolated populations could have an extinction probability much smaller than that of a species with a single population.

In most cases, however, local populations are close enough so that they sometimes experience similar environments and individuals may occasionally move among populations. Therefore their extinction probabilities may not be independent. In these cases, spatial correlation of environmental variation and dispersal are extremely important in estimating the extinction probability of a species. These are the two factors that we shall emphasize in this section.

5.1.1 Fragmentation

Loss of habitat often results in discontinuities in the distribution of the remaining habitat. Any population that inhabited the original area will be reduced to a smaller total size and also will be divided into multiple populations. Further fragmentation results in a decrease in the average size of habitat patches, an increase in the average distance between them (probably resulting in a decrease in dispersal rate), a decrease in the connectivity of patches, and in an increase in the edge:size ratio. Such changes may be exacerbated by changes in the environment throughout the landscape such as changes in fluxes of wind, water, radiation and nutrients (Saunders et al., 1991).

For many species, edge effects are an important component of patch dynamics that amplify the negative impacts of fragmentation (see, for example, Lovejoy et al., 1986). Changes in the environmental conditions at the edge of a patch (e.g. temperature, humidity, wind, light and disturbances such as fire, grazing, weed invasion, etc.) decrease the effective size of the patch for those species that inhabit the original habitat. They also provide conditions conducive to species that inhabit disturbed habitats. Some species may favour habitat edges, and the term edge effect is also used to describe an increase in species diversity in places where two habitats meet.

Both natural heterogeneity and fragmentation can occur at a variety of spatial scales, and this has important implications for the type of modelling that is most appropriate for the species studied and questions asked.

Consider a forest of 10 000 ha and a species (say, a predator) that

Fragmentation
The spotted owl (*Strix occidentalis*) has been listed as a 'bird species of special concern' since 1978 by the California Department of Fish and Game (Gould, 1985). In California there are three geographically distinct populations that belong to two subspecies. The northern spotted owl (*S. o. caurina*) occurs in the Pacific north-west. This subspecies was listed as threatened by the US Fish and Wildlife Service in 1990. The other two populations are considered to be a different subspecies, the California spotted owl (*S. o. occidentalis*).

An important factor in the recent decline of northern spotted owl populations is habitat loss, mostly due to the logging of old-growth forests on which the spotted owl is dependent. Habitat loss resulted not only in a decrease in the carrying capacity of the environment, but also in the fragmentation of the remaining habitat, causing populations to be restricted to small, relatively isolated patches of habitat. Even in regions where fragmentation is not as common as in northern California and Oregon, the natural heterogeneity of the environment restricts the species to distinct, relatively isolated habitat patches. Shaffer (1985) and other researchers have suggested that the case of the spotted owl fits the metapopulation concept, and development of metapopulation models will help the effective utilization of field data.

requires about 10 ha of forest habitat per pair. This forest will support an average of 1000 pairs. There are many ways in which the habitat could be fragmented. For example, habitat loss and fragmentation may reduce this forest to three fragments of 1000 ha each. The resulting metapopulation supports about 300 pairs, ignoring edge effects. This may be called a case of large-scale fragmentation, where a single population is split into several, but still relatively large, viable populations.

Now consider a second case, where the original habitat is reduced to 3000 fragments of about 1 ha each. The consequences of this small-scale fragmentation will be much different, although the total remaining habitat is the same as in the previous case. Since none of the fragments can support a pair (let alone a population), this system probably cannot be called a metapopulation. If the fragments are uniformly distributed, and if the individuals can easily move among fragments, it may be more appropriate to consider this system as a single population, which has 30% of the carrying capacity of the original habitat.

If the spatial scale of the fragmentation is somewhere between these two cases, or if in the second case the fragments are clumped, then the decision about whether to model the system as a single population with reduced carrying capacity, or as a metapopulation, will not be as clear. It will depend on the details of the fragmentation and on aspects of the biology of the species, such as the dispersal rate.

In the first case of fragmentation, the demographic characteristics of the three populations (growth rate, survivorship, fecundity) may be the same as those of the original population. In the second case, however, the large, diffuse population will probably have lower survival and fecundity rates, since the animals will have to travel among fragments to gather food, and this may decrease their efficiency and make them more vulnerable to predators. Edge effects, whatever these entail, will be much stronger in the second case.

The effect of fragmentation on greater gliders and southern brown bandicoots

Southern brown bandicoots (*Isoodon obesulus*) and greater gliders (*Petauroides volans*) are marsupials that inhabit the forests of eastern Australia. Possingham and Noble (1991) and Possingham *et al.* (1991a,b) used structured simulation models of the kind we describe below to evaluate the effects of disturbance at different spatial scales on the viability of populations of these species. Studies of this kind are important because they provide a qualitative guide to the likely effects of different management practices. For example, the size and distribution of areas clear-felled for timber production may be manipulated to minimize the detrimental impacts on these species, and the relative importance of the loss of different patches may be evaluated.

Different kinds of statistics may be derived from such analyses to answer particular questions. For example, they analysed the added risk to the southern brown bandicoot population that resulted from the removal of alternative patches of habitat (Figure 5.1). To address a slightly different question, they evaluated the size of a patch necessary to maintain a viable population of the greater glider (Figure 5.2).

Finally, note that the scale of fragmentation that is relevant for metapopulation modelling is not absolute, but is relative to the biology of the species. For a small animal that has a home range of only a few square metres, both cases discussed above represent metapopulations, although at very different scales.

In this spirit, Hanski and Gilpin (1991) introduced terms for three spatial scales that are defined with reference to the biology of the species. The **local** scale is the scale at which individuals move and interact with each other in the course of routine breeding and feeding. The **metapopulation** scale is the scale at which individuals frequently move from

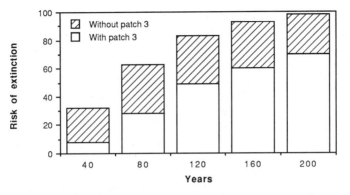

Figure 5.1 The probability of extinction of a metapopulation of the southern brown bandicoot in the Deep Creek area in South Australia. The added risk faced by the metapopulation due to the removal of a single patch (patch 3) from a system of five populations is represented by the hatched area. (After Possingham *et al.*, 1991a.)

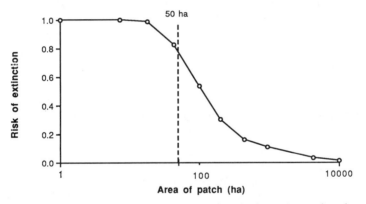

Figure 5.2 The effect of the size of an undisturbed patch on the chance of extinction of a population of greater gliders. Patches less than about 50 ha are of relatively little conservation value for this species. (After Possingham and Noble, 1991.)

one place (population) to another, typically across low-quality habitat. The **geographical** scale is the scale of the entire geographic range of a species, and across which individuals have little chance of moving.

5.1.2 Correlation of environmental variation among populations

Spatial correlation, or autocorrelation, measures of the amount of association between measurements of a variable made at points at a given distance apart. We could measure a single variable (say, rainfall) at intervals of 1 km across the landscape. The correlation coefficient between pairs of values recorded 1 km apart will tell us the amount of association between measurements of rainfall at points separated by 1 km. A correlation coefficient of unity indicates that the same pattern of rainfall (although not necessarily the same absolute amount) has been recorded everywhere in the landscape. A correlation coefficient of zero tells us that even if we know the rainfall at one point in the landscape, we cannot predict anything about the rainfall at any other point.

If two populations are close to each other geographically, they will often experience similar environmental patterns – for example, the same sequence of years with good and bad weather. This will result in a positive correlation between the vital rates of the two populations inducing a correlation between the population trajectories. Good rains in one habitat patch will mean there probably will be good rains in the other habitat patch. If rain limits reproduction, the changes in the sizes of the two populations will be positively correlated. If one population decreases (or becomes extinct) in a particular year, the other population will be more likely to decline in the same year.

In the extreme case where the populations are very close, this correlation may be perfect and they may become extinct at the same time. If we think of the two patches of habitat occupied by these populations in very general terms, they are either occupied or extinct in a given year (we ignore the effect of population size for a moment). Considering them as a single population will not be any different from considering them separately, in terms of the impact of environmental variation on extinction risk. In this case, if any number of populations has an extinction risk of E, the extinction probability of all the populations will also be E:

$$P = E. \tag{5.1}$$

At the other extreme, when the environmental variations that the populations experience are completely dissimilar, the total extinction probability will be a product of the probabilities as described above. For an n-population system,

$$P = E^n. \tag{5.2}$$

This formula assumes that there is no dispersal among populations. In the case of perfect correlation (Equation (5.1)), the rate of dispersal does not change the result. Hence, if we ignore other factors, particularly dispersal rates, and assume that geographic proximity means similarity of environmental variations, a species with local populations close together will have a higher risk of extinction than a species with local populations farther apart. This discussion can be made more general by considering ecological distance rather than geographic distance. Two populations on the opposite slopes of the same hill may be geographically close, but the population facing north may experience very different ecological conditions from the one facing south.

The difference between Equations (5.1) and (5.2) can be very important from the point of view of species conservation. To demonstrate this, consider a metapopulation of five local populations. Suppose that each population has an extinction probability of 1% within the next ten years. For this metapopulation, Equation (5.1) will give a total extinction risk of

$$P = E = 0.01$$

whereas Equation (5.2) will give a risk of

$$P = E^n = (0.01)^5 = 10^{-10}$$

which is many orders of magnitude smaller than the previous value. In most cases, the similarity of environmental factors (hence the correlation among extinction events) will be somewhere between these two extremes. If the models used to study metapopulations do not incorporate correlations, one needs to assume either of these two extremes. The assumption of perfect correlation (Equation (5.1)) will overestimate the extinction risk, whereas the assumption of independent environments (Equation (5.2)) will underestimate it.

It is easy to imagine how this underestimation may result in the needless extinction of a species: since the risks seem to be low, protective measures may not be taken. It is also possible, however, that the overestimation of risk could also lead to an avoidable extinction. Suppose that your job is to design nature reserves for a species. One option is the metapopulation described above, and another is a single patch in which you estimate the extinction risk to be, say, 0.001. If you use a model that assumes perfect correlation, you may select the single patch because it seems to result in a lower risk for the species. In fact, the extinction risk in the metapopulation may be lower (much lower, if you believe Equation (5.2)).

Den Boer (1968) alluded to this aspect of metapopulation dynamics when he noted that, when fluctuations are spread over a number of separate populations, the overall risk faced by the metapopulation is reduced. The risk is less because there is a degree of independence in the environmental fluctuations faced by different populations, so that fluctuations in population sizes are to some extent asynchronous (Figure 5.3). Levins (1969) made use of the same observation in recommending control measures for insect pests. In den Boer's (1981) study, populations of *Calathus melanocephalus* experience fluctuations that are much more highly correlated than those experienced by populations of *Pterostichus versicolor*. As a result, the *C. melanocephalus* metapopulation faces a much higher risk of extinction. Variation in the sizes of populations of *P. versicolor* may be less strongly correlated with one another through time because the species is less sensitive to correlated environmental conditions, or because it has better dispersal capabilities. To assess these

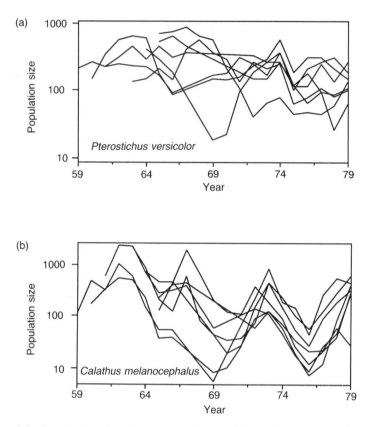

Figure 5.3 Population abundances in metapopulations of two carabid species. (After den Boer, 1981; see Hanski, 1991.)

risks correctly, a model is required that incorporates the correlation of environmental factors among populations. Note that we did not consider dispersal among populations, which makes things even more complicated. We shall discuss this factor next.

5.1.3   Dispersal patterns

If local populations do not exchange individuals, populations that become extinct will not have a chance to recover from extinction. Dispersal refers to the movement of individuals among spatially separate patches of habitat and it includes all immigration and emigration events. Dispersal among local populations will decrease the extinction risk for a species, if the patches where local populations become extinct can be recolonized by individuals from other populations. Dispersal mechanisms act to reduce the detrimental effects of correlated environmental variation in different

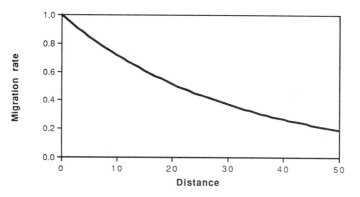

Figure 5.4 Dispersal rate as a negative exponential function of the distance between source and target populations. The curve is generated from Equation (5.3) using a value for *c* of 30.

habitat patches. We shall explore the interactions between these factors in several examples below and shall discuss them in detail in Section 5.1.4.

The rate of dispersal is mostly determined by the biology of the species including such things as its motility, propensity of juveniles to disperse, the mode of seed dispersal, and so on. These factors will determine the speed and ease with which a species explores and is able to colonize suitable habitat.

The dispersal rates between different populations of the same species may also differ drastically, depending on the population-specific characteristics. Factors that may influence dispersal include the distance of a population from the others, the surrounding landscape, and the density of the population (resulting in density-dependent dispersal). Dispersal from each local population to all others may not be possible, so the topology of the migratory pathways will strongly influence the overall extinction risk. For example, freshwater fish are able to recolonize tributaries within the same river system, but are not able to recolonize other river systems. The exact effect of dispersal on species extinction will depend on the characteristics of the dispersal patterns. We shall now discuss factors that are the important determinants of these differences in the rate of dispersal.

(a) *Distance*

Dispersal occurs at a higher rate when the populations are geographically close. The relation between dispersal rate and distance can be described by a negative exponential function (see Figure 5.4 for an example). Wolfenbarger (1946) studied the dispersion of 47 species, all of which showed a monotonically declining rate of dispersal with distance similar to the negative exponential function. Kitching (1971) showed that a simulation model of dispersion also gave similar results. This function of dispersal can be written as

$$m = e^{-d/c}, \tag{5.3}$$

where m is the rate of dispersal (expressed as the proportion of the current source population leaving per unit time), d is the distance between the source and target patches, and c is a constant representing the average distance a migrant travels per unit time. The reciprocal $1/c$ represents the death rate of migrants per unit distance. A similar formulation is (Gilpin and Diamond, 1976):

$$m = \frac{e^{-d^2/c}}{d}. \tag{5.4}$$

Occupancy patterns determined by dispersal
Fritz (1979) studied spruce grouse (*Canachites canadensis*) populations in the Adirondack mountains in New York (USA). The metapopulation consisted of 25 occupied and 7 unoccupied patches. Occupied patches were significantly larger and were closer to the nearest occupied patch than unoccupied patches. This observation implies that distance-dependent dispersal is an important factor in determining the occupancy pattern for this species.

Other factors besides distance may also affect the topology of migratory connections in a metapopulation. Even if two patches are geographically close, the unsuitable habitat or geographic barriers that separate them may make the dispersal rates small. On the other hand, habitat 'corridors' may facilitate dispersal and therefore connect distant patches.

Corridors
Patches of woodland habitat in farmlands in Ottawa (Canada) provide an example of the importance of corridors. Fence-rows run through the farmland surrounding the patches of woodland. The fence-rows are wooded, thus providing habitat corridors that connect the woodland patches. As a result, the patches are not isolated; small mammals such as chipmunks (*Tamias striatus*) and white-footed mice (*Peromyscus leucopus*) migrate using these corridors and recolonize empty patches (Middleton and Merriam, 1981, 1983; Wegner and Merriam, 1979). Henderson *et al.* (1985) studied the dynamics of this metapopulation experimentally, by clearing two patches of all chipmunks (causing two local extinctions).

For both experimental and natural occurrences of local extinctions, recolonization of the vacated patches succeeded in 24–53 days. All colonists were non-reproductive individuals who were mostly less than one year old. More than 90% of the movements between patches were made using wooded corridors, although non-wooded gaps of 20–60 m in these corridors were crossed commonly, and gaps up to 460 m were crossed occasionally. Merriam (1991) speculates that without the corridors the chances of successful recolonization of empty patches would have been reduced, resulting in elevated risks of extinction for the entire metapopulation.

Dispersal can occur at different rates in two directions between two populations. For instance, local populations of an aquatic species along a river may have dispersal mostly or only in the direction of the waterflow, i.e. downstream, but not upstream. This case can be modelled using the dispersal formula given above by specifying two different constants c in Equation (5.3), one of which (for upstream dispersal) can be either zero (yielding $m = 0$), or much smaller than the other.

(b)   *Population density*

Density-dependent dispersal can be an important aspect of the ecology of species. If organisms have a higher tendency to emigrate from their population under crowded conditions, this will yield not only a larger number, but also a greater proportion of individuals leaving their population as density increases, resulting in dispersal rates that are increasing functions of abundance. A similar effect can also occur in plant populations if, for example, a high density causes an aggregation of frugivorous organisms which help dispersal, resulting in a higher proportion of seeds dispersed compared to smaller populations.

Under such density-dependent mechanisms, the rate of dispersal (and consequently the probability of recolonization of other patches of habitat) is a function of the density of the source population. Another type of density dependence in dispersal rates occurs when the organisms have a higher tendency to emigrate from smaller populations, resulting in a stepping stone effect.

Stepping stone dispersal
Field voles (*Microtus agrestis*) in the Tvärminne archipelago, Finland, can migrate between islands by swimming (Pokki, 1981). The rates of dispersion from different sized islands in this archipelago provide an example of stepping stone dispersal. Pokki (1981) calculated dispersal rates by marking voles on several islands, and found that 8.5% of the voles marked on small islands or skerries

(<1 ha) dispersed, whereas this rate was 2.5% from medium sized (1.1–5.0 ha) islands, and 0.6% from larger islands (>5 ha). Higher emigration from smaller islands suggests that these islands are used as stepping stones by the voles.

The mechanism of density dependence discussed above assumes that at the time the organisms migrate from one patch to the other, they can only perceive the density of the patch they are leaving, not the density of the patch they are going to. However, there is also some evidence (Smith and Peacock, 1990) that emigration rates can also depend on the density of the target population. According to this hypothesis, the animals use the density of the target population as an indicator of the environmental quality of the target patch, and if the density is very low they do not settle in that patch, but continue on to another one. Such patches with low habitat quality or small size may serve as stepping stones, since the organisms will spend less time in patches with low population densities. These patches will be used by the organisms only temporarily before they continue their migration.

(c)   *Age, sex and genetic structure of the populations*

Dispersal rates can be age- or sex-specific, such as when only immature individuals or only young males disperse to other habitats. The shrew (*Crocidula russula*) used as an example in Chapter 3 exhibits this kind of dispersal pattern. Most juvenile shrews disperse from their parents' home range in the same year they are born, recolonizing patches left empty from local extinctions during the previous winter, or invading already occupied territories (Cantoni and Vogel, 1989).

If the age and sex structure of the subpopulations are not very different from each other, age- or sex-specific dispersal may not have a significant effect on metapopulation dynamics. The characteristics of individuals that disperse from populations may be different from the characteristics of individuals in populations that receive them (Hansson, 1991). If, for example, different populations have different age structures (different proportions of individuals in each age class; see Chapter 4) then the arrival of immigrants will push a population away from its stable age distribution. This may induce population fluctuations. However, even when the age and sex structures are similar, age- and sex-specific dispersal may have an important effect on the genetic structure of the subpopulations (see Chapter 6).

It is very difficult to draw generalizations about the effectiveness of dispersal in reducing extinction probabilities. Many dynamic and behavioural characteristics are unique to individual metapopulations.

Human-mediated dispersal can have significant effects on extinction

Age-specific dispersal
Gill (1978) studied the dynamics of a red-spotted newt (*Notoph-thalmus viridescens*) metapopulation in Virginia (USA). This meta-population of newts exists in 12 artificial mountain ponds that range from 10 to 14 m in diameter. Gill found that only some of the populations were productive; in others no juveniles were produced, but they continued to support populations as a result of colonization from the other ponds. Hence the non-productive ponds are popu-lation sinks that do not contribute to the metapopulation growth. However, there was also some turnover of these sinks; that is, the ponds that were productive changed from year to year.

Another important characteristic in this metapopulation was the age dependence of dispersal. Adult red-spotted newts showed a high site-fidelity, every year returning to breed in the same pond that they first encountered as juveniles. Dispersal between ponds (hence recolonization of sinks) was entirely by juvenile dispersal.

probabilities. Translocation of individuals from high density populations to empty or low-density patches may decrease extinction risks. Re-introduction of captive-bred individuals to empty patches that were previously inhabited by the same species may also be effective in some circumstances (Griffith *et al.*, 1989).

Dispersal and correlation patterns
The bay checkerspot butterfly (*Euphydryas editha bayensis*) is a threatened species that exists in two metapopulations in California (Ehrlich *et al.*, 1975; Murphy *et al.*, 1990). It has a low dispersal rate (Ehrlich, 1961), lives in a patchily distributed habitat, and its local populations are prone to frequent local extinctions (Ehrlich *et al.*, 1980). Harrison *et al.* (1988) studied the ecology of one of its metapopulations, which consists of one large (2000 ha) and about 60 smaller (0.1–250 ha) patches of suitable habitat. The carrying capacity of the patches in this metapopulation are determined by their area, topography and resource abundance. The distribution of occupied and unoccupied patches was predicted by Harrison *et al.* who used a simulation model with dispersal rate expressed as a negative exponential function of the distance from the main popu-lation, with $c = 1.7$ km.

The authors concluded that the rate of dispersal from the main population is an important factor determining the pattern of patch occupancy, especially after major disturbances. One such distur-bance was a severe drought which caused the extinction of most subpopulations. In addition, the changes in the abundance of two populations of checkerspot butterfly elsewhere were correlated with rainfall patterns, indicating that the fluctuations in the two populations were caused by similar environmental factors. This is an example of correlated environments experienced by subpopulations: a large-scale fluctuation in environmental conditions may affect most or all of the metapopulation.

Harrison *et al.* concluded that the spatial extent of the metapopulation was primarily limited by the slow dispersal rate of the species. The rate is so slow that not all the populations in patches that become extinct after major droughts are recolonized. Major droughts occur approximately every 50 years. They also stated that other environmental disturbances may increase the frequency of local extinctions, decreasing the spatial extent of the metapopulation.

Another factor that may affect the spatial extent of the metapopulation is the effect of the stepping stone phenomenon on colonization rates. It may result in faster dispersal compared to dispersal from the main population alone, thereby increasing the spatial extent by facilitating colonization of the most distant patches. Although there is some evidence for stepping stone effects in the form of a higher per capita emigration rate when the population size is small, a simulation by Harrison *et al.* showed that small populations should have emigration rates orders of magnitude higher than larger populations in order to significantly change the occupancy pattern of the metapopulation.

5.1.4 Interaction between dispersal and correlation of environmental variation

The extinction risk of a species will depend on the nature of, and the interrelationships among, all the factors discussed above. A species that lives in local populations that are far away from each other may have largely independent extinction probabilities for each of its local populations, and will perhaps have a lower overall extinction risk. On the other hand, dispersal rates will probably be lower for such a species compared to a species with closer local populations. Hence, in some cases there may be a trade-off between similar environments (and, consequently, correlated extinctions) and low dispersal rates for distant populations.

Another complexity is the interaction between these two factors. Interaction refers to how much the effect of one factor depends on another factor. In this case, the degree to which dispersal improves the chances of survival of a species depends on the similarity of the environments that the populations experience. For example, in the extreme case of perfect correlation of environments, populations may become extinct at the same time; hence whether there is dispersal or not before this time will not change the extinction risk. Dispersal will decrease this risk only if the populations become extinct at different time periods so that the extinct patches have a chance of being recolonized by migrants from extant populations.

Another way in which dispersal and correlation interact was illustrated by Forney and Gilpin (1989) who designed an experimental system of *Drosophila* populations and demonstrated the effect that dispersal between subpopulations has on the probability of extinction. Two isolated populations had a higher extinction risk than two populations connected by a migration corridor, which in turn had a higher risk than a single large population. As they pointed out, their separated populations had correlated extinctions, which probably increased the extinction probabilities of both the two separated and two connected populations.

Their results also suggest another interesting form of interaction between the two factors: increased dispersal between two populations may increase the correlation of their extinctions since migrating individuals can transport diseases from one population to another. In some cases this factor may offset the benefit of increased dispersal.

Similar interactions and trade-offs exist among other components of spatial structure. One example relates to the argument about whether a single large or several small reserves of the same total size provide better protection for a species against extinction. A single population will not benefit from uncorrelated environmental fluctuations; if it becomes extinct it cannot be recolonized from others. A metapopulation of several populations may have a lower extinction risk, if the rate of dispersal is high enough and the degree of spatial correlation of environments is low enough. These threshold levels of dispersal and correlation that make several small reserves better than a single one depends on many other factors, such as the configuration, size and number of populations. For example, a single large population will suffer less demographic stochasticity compared to a species with several small populations.

It is important to note that the trade-off between large and small reserves that we are analysing only applies to the case where the total area of a single large reserve is more or less equal to the total area of small reserves. A fragmented habitat that has several small patches certainly contains a smaller (and a more extinction prone) total population compared to the original, non-fragmented habitat. This is because as a result of fragmentation, usually not only the total area of habitat is drastically reduced, but also the movement of individuals (migration, dispersal) is restricted. Furthermore, the resulting pieces of fragmented habitat are generally no more independent of each other than they were before fragmentation (because they are at the same locations in the environment). However, certain complicating factors (such as slower spread of fires, disease outbreaks, etc.) can potentially make the fragments more independent than parts of a single large habitat. For example, disease spread in the *Drosophila* example above resulted in correlations between the environmental factors experienced by connected populations.

The above discussion of reserve design also assumes that the size of the patches that are selected remains the same, or at least, the decline in size does not depend on whether one large or several small patches have been selected. In fact, because of edge effects, the original habitat in smaller patches may erode faster than in larger patches. All these interactions and complexities make it impossible to assess species extinction risks or address questions about reserve design solely by means of intuition. In the next section we shall discuss metapopulation models that have been developed to address these questions.

5.2 OCCUPANCY MODELS

The simplest model for the dynamics of metapopulations is that underlying the theory of island biogeography, a concept we outlined in Chapter 1. The variable of interest in this approach is the equilibrium number of species in a patch of habitat. Thus, island biogeography generally treats a community (or a guild) of different species as a whole. The models discussed in this chapter consider each species separately, and try to estimate their risks of extinction one by one. The model for island biogeography is a special case of a more general class of models called occupancy models, first developed by Levins (1969, 1970).

In occupancy models, the variable of interest is the proportion of patches (habitat fragments, islands, etc.) occupied by the species. In population dynamic models, the variable is the size of the population in each patch and in the entire metapopulation. Since these two types of models are based on different variables, the questions they address, as well as the parameters they require, are quite different. In the following sections, we shall discuss the basic structure, assumptions and limitations of these two approaches in detail. Although for the purposes of extinction risk assessment we favour the more complex models that are based on species-specific information, the choice depends on the availability of data. Each of these models has a place in shaping our basic understanding of metapopulation dynamics and in helping decision making in the face of limited information.

The metapopulation models discussed below predict the equilibrium proportion of patches that are occupied by a species based on the probability of extinction and recolonization of populations in a single patch. In these models, each patch is considered to be either occupied or empty, thus ignoring the dynamics of population growth in each patch. We shall use the term *occupancy model* for this kind of metapopulation model.

5.2.1 Levins' model

As the interest in conservation biology shifted towards processes from patterns, the island biogeographic approach was replaced by the meta-population approach as the paradigm in conservation biology (Hanski, 1989). Levins (1970) was the first to use the term metapopulation, and he developed the first model for metapopulation dynamics.

A common characteristic of occupancy models is that they have extinc-tion probabilities as parameters. In the population models of previous chapters, extinction probability was often one of the output variables, not a parameter. The parameters included such things as the population's vital rates (survivorship and fecundity). Most of the early metapopulation models were occupancy models and their main parameters were local extinction and recolonization probabilities.

The metapopulation model of Levins (1970) became the basis of many other modelling efforts. In his model the variable of interest is the proportion of patches that are occupied. The rate of change in the proportion of occupied patches dp/dt is calculated as increase due to colonizations minus decrease due to extinctions:

$$dp/dt = mp(1 - p) - Ep \tag{5.5}$$

where m is the colonization parameter defined as the probability of dispersal from an occupied patch to any other patch per unit time. The parameter E is the probability of extinction of a given local population in a unit time interval. Colonization is the first term on the right-hand side of Equation (5.5). It is assumed to be proportional to the product of occupied patches p and unoccupied patches $1 - p$, and extinction (the second, negative term, Ep) is proportional to the number of occupied patches. The model has a positive equilibrium p^* if $m > E$:

$$p^* = 1 - \frac{E}{m}. \tag{5.6}$$

Thus, the metapopulation becomes extinct if the probability of ex-tinction of local populations exceeds the probability of colonization. This means that for the metapopulation to persist, the total number of col-onizations per unit time must exceed the total number of extinctions when the proportion of occupied patches is small; as p gets larger, the two rates become closer and, of course, at p^* they are equal. Since the model is not stochastic, it cannot make predictions such as the risk of extinction, or mean time to extinction. As long as $m > E$ and there is at least one occupied patch initially, the metapopulation persists indefinitely.

Hanski and Gilpin (1991) show that a single-species version of the MacArthur–Wilson (1967) model of island biogeography corresponds to an occupancy model in which there is a continental source population that never becomes extinct. It has a positive equilibrium at the point

$p^* = m/(m + E)$, so there will always be an inhabited patch in a system that has any turnover whatsoever. They show further that Equation (5.5) is structurally equivalent to the logistic equation (Chapter 3), where $(m - E)$ is the rate of increase of p when p is small, and $(1 - E/m)$ is equivalent to the local carrying capacity.

5.2.2 Assumptions of Levins' model

The model of Levins and some other models based on the same structure have a number of important assumptions.

(a) *No local population dynamics* The model assumes that each population is either extinct (i.e. population size is 0) or it is extant (e.g. the population is at its carrying capacity, K). The sizes of these populations are ignored. A recolonized patch goes from empty to a population at the carrying capacity in a single time step. This assumption is only valid when the time scale of local population dynamics is much faster than the time scale of regional events such as recolonization of extinct patches and the extinction of occupied patches. Metapopulation extinction probabilities will be underestimated if a population requires several time steps to reach the carrying capacity and the local extinction probability is related to population size rather than the carrying capacity of a patch. An important corollary of this assumption is that the local extinction probability is constant, whereas in real populations it depends on the population size at that time interval. Another corollary is that local population dynamics are not affected by immigration and emigration.

(b) *Infinite number of equal patches* All local populations are assumed to have the same probability of becoming extinct per unit time. This may be the case if all the populations are identical (e.g. they have the same equilibrium density or carrying capacity, the same habitat quality and the same population dynamics). In addition, p (the proportion of patches that are occupied) is a continuous variable. This implies that there are an infinite number of patches.

(c) *Independent extinctions* Local populations are assumed to become extinct independently of each other, i.e. the extinction events in neighbouring patches have zero correlation. This is possible only if patches experience independent environmental variation. If correlations of environmental factors are distance-dependent, this is a reasonable assumption only in cases where populations are relatively far from each other.

(d) *Equal probability of dispersal* Dispersal and recolonization events are assumed to be equally likely from any patch to any other. This may be a reasonable assumption in some systems where the rate of dispersal does not depend on the distance between patches. Alternatively, if dispersal is distance-dependent, it may be a reasonable assumption if the distance between patches is more or less the same. This, in turn, is possible if there are a small number of patches (but see (b) above), or the patches are so close to each other that the distance between the two nearest patches is not much shorter than the distance between the two most distant patches (Hanski, 1983). If this is not the case and dispersal occurs mostly among neighbouring patches, then the model overestimates colonization rate when the proportion of occupied patches is small (Nisbet and Gurney, 1982).

Looking at assumptions (c) and (d), it is obvious that this model ignores various components of the spatial arrangement of populations. In fact, if both dispersal and correlation are distance-dependent, these assumptions may be contradictory: (c) implies that the local populations are very far away; (d) implies that they are very close to one another. However, this model can be considered to be a first approximation, which points to the important (although perhaps obvious) relation between rate of patch extinctions, rate of recolonization of extinct patches and the persistence of the metapopulation. Verboom *et al.* (1991) evaluated the adequacy of Levins' model and found it to be an adequate approximation of a more complex model incorporating local population dynamics for a metapopulation of the European badger (*Meles meles*).

5.2.3   Generalizations of the model

Several authors have modified the metapopulation model of Levins to make it more general and realistic (see Hanski, 1991). In addition, multispecies interactions have been modelled using this approach as a basis (e.g. see Slatkin, 1974, and Hanski, 1983, for competition models, and Taylor, 1988, 1990, for predation models). For single species dynamics, Nisbet and Gurney (1982) incorporated the effect of a small number of populations. This step is comparable to accounting for demographic stochasticity arising from small number of individuals in a population. They calculated an approximation for the mean lifetime of a metapopulation, which increases exponentially with the number of patches.

Hanski (1983) incorporated the effect of population size. He argued that average local abundance may be positively correlated with the proportion of occupied patches, p (due to a larger number of immigrants from a larger number of occupied patches), and negatively correlated with E (small populations are in general more extinction prone); thus p will be negatively correlated with E. He rewrote the model as

$$dp/dt = mp(1 - p) - E(1 - wp)p \qquad (5.7)$$

where w is a parameter describing the negative correlation between p and E. When p is small, the local extinction probability is E, but when p is large, the extinction probability is reduced by a factor of $(1 - wp)$. At the cost of introducing a new parameter w, which may be difficult to estimate, this model takes into account some of the effects that the Levins' model ignores. (The problem of estimating the other parameters E and m, which are common to all occupancy models, will be discussed below.)

Another approach is to incorporate population dynamics by modelling the age (or some other property) of patches. This method is mathematically equivalent to von Foerster single-population models that incorporate age structure (Metz and Diekmann, 1986). Hastings and Wolin (1989) developed such a model in which the populations grow deterministically until they are reduced by a disaster. As in Levins' model, they assumed a large number of populations with dynamics independent of dispersal, and ignored spatial structure. Gyllenberg and Hanski (1990) developed a similar model which takes into account the effect of dispersal on local population dynamics. Both of these models assume that extinctions in different patches are independent.

Harrison and Quinn (1989; see also Gilpin, 1990) modified Levins' model to incorporate correlated extinctions in different patches. To do this, they made a few assumptions.

1. There are two distinct extinction probabilities, corresponding to 'good' and 'bad' time steps (say years). In a good year, the extinction probability $E = \mu - \delta$, in a bad year $E = \mu + \delta$, where δ is a random number with mean equal to zero.
2. The two types of time steps, good and bad, are equally probable.
3. The random component δ of the extinction probability is further divided into two components. One component, of magnitude $c\delta$, is experienced by local populations in all patches, the other component, $(1 - c)\delta$, varies independently among local populations. Thus c is a measure of the correlation of extinction probabilities among local populations in different patches.

This model adds an important factor, environmental correlation, to Levins' model. Although this factor is described by a single coefficient (thus all pairs of patches have the same correlation regardless of the distance between them), its effect on the occupancy rate in the metapopulation can be demonstrated with this approach. The results of this model are expressed in terms of the mean time to extinction by Harrison and Quinn (1989). They found that in the case of uncorrelated environments, this variable is roughly proportional to $[m/E]^n$, where n is the number of occupied patches. For the case of correlated environments, they concluded that correlation is important only in large metapopulations in which the mean and variance of the extinction rate are high.

In summary, despite their restrictions, these generalizations show that the occupancy models are flexible. They can be useful in cases where the amount of information is limited and the aim is to understand general patterns rather than make predictions for specific metapopulations. Occupancy models are in general similar to each other in the types of predictions they can make. Therefore, we shall not investigate each of the variations of these models separately. Instead, we shall analyse a more general occupancy model in the next section. This analysis will give a general idea about the use of occupancy models, and will also highlight their limitations.

5.2.4 An occupancy model with spatial structure

The occupancy model that will be illustrated here is a new approach that further generalizes the occupancy idea. It allows a specific extinction probability for each patch, as well as a specific correlation and a specific recolonization probability for each pair of patches in a metapopulation (Akçakaya and Ginzburg, 1991).

The parameters for a given patch are estimated from a model of the dynamics of a population from a single locality. This approach solves the problem of species viability in two steps, first by generating local extinction parameters from a complex model of a single population and then analysing the total (i.e. metapopulation) problem in much cruder terms. In this model the local extinction probabilities are not the same for all patches, but are patch-specific; thus the local populations are not assumed to be identical. In addition, the model has full spatial structure, with a specific correlation coefficient and a specific dispersal rate for each pair of populations. With this model, it is possible to compute the risk of extinction for any subset of populations, in addition to the risk of extinction of the entire metapopulation.

(a) *Variables and parameters of the model*

The metapopulation model is based on a Markov process of transition between states, each of which represents the occupancy status of local populations. Each state is described by a time-dependent variable that represents the probability that a particular combination of populations remains extant to the end of time t. Consequently, the number of states is equal to 2^n, where n is the number of populations. The model is described here for the case of two populations, in which case there are four variables in the model. The four variables that describe the probability of each of the four states are

$S_{11}(t)$ Both populations are extant at time t
$S_{10}(t)$ Population 1 is extant, and population 2 is extinct at time t
$S_{01}(t)$ Population 2 is extant, and population 1 is extinct at time t
$S_{00}(t)$ Both populations are extinct at time t

Table 5.1   Conditional probabilities of extinction in a metapopulation composed of two populations.

Population 2	Population 1		
	Extinction	*Survival*	*Total*
Extinction	E_{00}	E_{10}	E_{*0}
Survival	E_{01}	E_{11}	E_{*1}
Total	E_{0*}	E_{1*}	1.0

These four states make up a vector of state variables (**S**) that is used to keep track of the metapopulation. At each time period, the transition from each of these states to the others is described by transition matrices. These transition matrices are based on two sets of parameters. The first set consists of conditional probabilities of population extinction in a given year (E_{ij}). These are the probabilities that a particular population becomes extinct or survives in a single time period, given that the other population becomes extinct or survives. The probabilities are defined in Table 5.1.

Many other formulations are possible. The model may have been developed so that a population could be in one of three states: extinct, extant and threatened, or extant and safe. Such a system would involve 3^n states, where n is the number of populations, and would be computationally much more difficult to solve than the two-state system we describe in detail here.

The column and row totals in Table 5.1 show the marginal probabilities. For example E_{0*} is the probability that population 1 becomes extinct in one time period, regardless of what happens to population 2. The set of four conditional probabilities describe the degree of dependence between the survival of the two populations. There are an infinite number of combinations of these conditional probabilities that will give the same marginal totals, each of which correspond to a particular degree of association or dependence between the two populations.

For example, for a species distributed in two patches, there may be a marginal probability of extinction of 0.2 for both populations. That is, each of the populations may face a 0.2 chance of extinction, if you ignore the state of affairs in the other population. The data in Table 5.2 represent three different combinations of conditional probabilities that give the same marginal probabilities.

Combination A represents complete dependence or association between the two populations. The marginal extinction probability of each population is equal to E_{00}, i.e. the extinction probability of each population is also the extinction of probability of the metapopulation. Thus

Table 5.2 Three examples of conditional probabilities of extinction for a metapopulation composed of two populations. In each case, the marginal probabilities are the same. Case A represents complete dependence, case B represents complete independence, and case C represents negative association.

A			B			C		
0.2	0	0.2	0.04	0.16	0.2	0	0.2	0.2
0	0.8	0.8	0.16	0.64	0.8	0.2	0.6	0.8
0.2	0.8	1.0	0.2	0.8	1.0	0.2	0.8	1.0

they can only become extinct together. This represents the case where patches are experiencing the same environmental conditions.

Combination B represents complete independence of the two populations. In this case the product of the diagonal elements (0.04 × 0.64 = 0.0256) is equal to the product of the off-diagonal elements (0.16 × 0.16 = 0.0256). The probability of both populations becoming extinct per unit time (0.04) is the product of the marginal probabilities of each population becoming extinct per unit time (0.2). It represents a situation in which environmentally induced variation in the two populations is uncorrelated.

The other extreme, negative association, is represented by the combination C, where E_{00} is zero, i.e. if one population becomes extinct, the other survives. The conditional probabilities for real populations will probably be somewhere between A and B, and will not extend to C.

A second set of parameters describes the dispersal between the two patches of habitat that results in successful recolonization of an empty patch. The parameters M_{ij} represent the probability of dispersal and recolonization from population i to population j per unit time. M_{ij} need not be equal to M_{ij}; the dispersal between two populations may be asymmetric.

(b) *Transition matrices*

To predict the occupancy of the patches in the metapopulation, the parameters defined above must be organized. In their work, Akçakaya and Ginzburg (1991) used a set of matrices. These are, in a sense, similar to the transition matrices such as the Leslie and Lefkovitch matrices discussed in Chapter 4.

The difference is that instead of describing transitions of individuals among age or stage classes, the matrices describe the transition from one pattern of occupancy to another: each element of a matrix represents the probability of transition from a state at time $t - 1$ (columns) to a state at time t (rows). The first matrix is based on the conditional probabilities of extinction and their marginal totals described above (Table 5.1) and

Table 5.3 The matrix (**E**) of transition probabilities (E) between states, for extinction.

To state S_{**} at time t	From state S_{**} at time $t-1$			
	S_{11}	S_{10}	S_{01}	S_{00}
S_{11}	E_{11}	0	0	0
S_{10}	E_{10}	E_{1*}	0	0
S_{01}	E_{01}	0	E_{*1}	0
S_{00}	E_{00}	$1 - E_{1*}$	$1 - E_{*1}$	0

Table 5.4 The matrix (**M**) of transition probabilities (M) between states, for migration.

To state S_{**} at time t	From state S_{**} at time $t-1$			
	S_{11}	S_{10}	S_{01}	S_{00}
S_{11}	1	M_{12}	M_{21}	0
S_{10}	0	$1 - M_{12}$	0	0
S_{01}	0	0	$1 - M_{21}$	0
S_{00}	0	0	0	1

represents the events that determine the extinction and survival of populations (excluding dispersal).

For example, if the species is in state S_{10} at time $t-1$ (population 1 extant, population 2 extinct), then in the next time step it will either stay at the same state (population 1 will survive) with probability E_{1*}, or it will go to state S_{00} (population 1 will become extinct) with probability $1 - E_{1*}$ ($= E_{0*}$). All possible transition paths for the case of two populations are described in Table 5.3.

The second matrix is based on recolonization probabilities, and has the same properties. For example if the species is in state S_{10} at time $t-1$ (population 1 extant, population 2 extinct), then in the next time step it will go either to state S_{11} (by recolonization from population 1 to population 2) with probability M_{12}, or it will stay in state S_{10} (i.e. no recolonization) with probability $1 - M_{12}$.

There are a number of important things to note about the transition matrices illustrated in Tables 5.3 and 5.4.

1. Since these are transition matrices, their columns add up to 1.
2. A species at state S_{00} (both populations extinct) will remain at that state.
3. Dispersal has no effect if the species is in state S_{11} (all populations extant).

4. Extinction events (Table 5.3) either decrease or have no effect on the number of extant populations.
5. Dispersal events (Table 5.4) either increase or have no effect on the number of extant populations.
6. If the probabilities of all dispersal events are zero, i.e. the populations are completely isolated from one another, the dispersal matrix is reduced to a unity matrix (with diagonal elements equal to 1 and off-diagonal elements equal to 0).

These conditions apply to transition matrices for the case of two populations (given above) as well as those for the case of three populations (which are given in Akçakaya and Ginzburg, 1991).

The transition matrices with their conditional probabilities summarize all we need to calculate the fate of a metapopulation. The next step is to perform the necessary calculations. The probability of each state, S_{**}, at time t is calculated by the multiplication of the vector of state variables at the previous time period, $S(t - 1)$, with the transition matrices E and M:

$$S(t) = EMS(t - 1). \tag{5.8}$$

The probability of extinction of the entire species is given by the state variable $S_{00}(t)$, which is the probability that both populations will be extinct at time t. The same analysis can also made with, say, three populations, in which case there are eight states and the state $S_{000}(t)$ gives the probability that all three populations (hence the species) will be extinct at time t.

5.2.5   Sources of uncertainty

Occupancy models ignore the details of the dynamics of local populations. The various components of genetic, demographic and environmental stochasticity are encapsulated in the single number that is the extinction probability of a population in a patch. However, if population size and spatial distribution are included in an occupancy model, stochasticity unique to metapopulations is modelled explicitly.

Hanski (1991) makes a useful distinction between two sources of uncertainty. **Immigration-extinction stochasticity** is the equivalent of demographic stochasticity in single population models. It is the variability in the total population size that occurs through random events in dispersal and the recolonization of empty habitat patches. Even in circumstances where dispersal from an occupied patch to an unoccupied one is possible, it may not occur. A metapopulation may become extinct because all the local populations become extinct at the same time, even if the analogous deterministic model predicts a stable, positive equilibrium proportion of occupied patches.

The other source Hanski (1991) notes is **regional stochasticity**, the equivalent of environmental stochasticity for single populations. It is the amount of region wide variation experienced by a metapopulation. In other words, it is the strength of the correlation between environments experienced by the various populations. High levels of regional stochasticity will lead to elevated species extinction risks.

5.2.6   Risk assessment with occupancy models

We shall demonstrate the use of occupancy models for extinction risk assessment with a specific example from an endangered species, the Mountain Gorilla (*Gorilla gorilla beringei*), using the detailed occupancy model described above. Other occupancy models that incorporate correlated extinctions (e.g. Harrison and Quinn, 1989; Gilpin, 1990) would make qualitatively similar predictions. Because of the simplifying assumptions of such occupancy models, and because of the lack of detailed information on this species, the results are useful only to illustrate the general effects of spatial factors on species extinction risks. They do not represent a thorough or complete analysis of the viability of mountain gorilla populations.

Akçakaya and Ginzburg (1991) used data on Mountain Gorilla populations, in the Virunga Mountains of south-west Uganda, which live in three populations more or less isolated from other gorilla populations. They used information given by Harcourt *et al.* (1981) and Webber and Vedder (1983) on these populations to calculate age-specific survivorship and fecundity values for five-year periods, and then estimated the annual probability of extinction for a single population using an age-structured, single-population simulation model.

They repeated this process, first dividing the initial number of individuals by 2 and then by 3 to estimate the annual probability of extinction for populations $\frac{1}{2}$ and $\frac{1}{3}$ the size of the first population. After this process of parameter estimation, they used the extinction probability for the $\frac{1}{2}$-size population in a 2-population occupancy model, and that for the $\frac{1}{3}$-size population in a 3-population occupancy model. The model worked in the way described in the previous section. They used it to calculate the species extinction probability for 500 times steps (years).

For natural populations, it is not always easy the know the true values for such things as the chance of successful recolonization of a patch vacated by a population. Even when these data are missing, it can be a very useful exercise to evaluate the importance of parameters such as the chances of recolonization. With the application of common sense and whatever qualitative biological observations are available, it is usually possible to guess at the largest and smallest values a parameter could have. For example, if people have been observing several populations for decades and have not seen one migrant pass between the patches in that time, the annual probability of successful recolonization is likely to be at the small end of the scale. Akçakaya and Ginzburg (1991) studied the effect of environmental correlation and migrations among two and three populations by using a series conditional probabilities representing a range of correlations (r) from 0 (independence) to 1 (complete dependence), and by using a range of dispersal/recolonization probabilities that ranged from 0 to 1% per year.

The results are interesting because they show that the extinction probability of the species in the Virunga Mountains is sensitive to factors for which the data are scarce: correlation of environmental variation among localities and the probability of recolonization among populations. The data on which the analysis is based come from a metapopulation for which there is insufficient information about population subdivisions and about the life history of the subpopulations. Therefore, the interpretation of results is necessarily an exploratory exercise aimed at a qualitative analysis of the trade-offs between various factors affecting metapopulation dynamics, rather than at quantitative predictions about the future of a specific population.

In an occupancy model, the probability of remaining extant for one year for a single population is $1 - E$, where E is the annual extinction probability. The probability for staying extant for t consecutive years is $(1 - E)^t$, and the probability of extinction at or before year t is $P(t) = 1 - (1 - E)^t$. $P(t)$ is identical for a single population or for any number of populations with the same E in perfectly correlated environments. Two curves in Figure 5.5 (a and b) give the extinction probabilities as a function of time for single populations of full and $\frac{1}{3}$-size respectively. Curve b is identical to the extinction risk for three populations each with $\frac{1}{3}$ size (the 3-population system) in perfectly correlated environments ($r = 1$).

When environmental correlation is such that extinction events are independent, and there is no dispersal, the probability that all n identical populations of a species will be extinct by the end of year t is $P(t)^n$, where $P(t)$ is the probability for a single population. Curve c in Figure 5.5 shows the species extinction probability in the case of the 3-population system with independent extinctions. Curves a and c, for correlations of 1 and 0, bound all the cases of intermediate correlation (with no recolonization) for the 3-population system. That is, if there is no dispersal between the gorilla populations, the true extinction curve will lie somewhere between these two.

The difference between the extinction probability curves of a large (full size) and a small ($\frac{1}{3}$ size) population is due to two factors. First, a small population is nearer the extinction threshold, so environmental fluctuations can make its decline to extinction much easier. Second, a small population is affected more by demographic stochasticity, which increases the year-to-year variation in population sizes. Other things being equal, small populations have higher extinction risks.

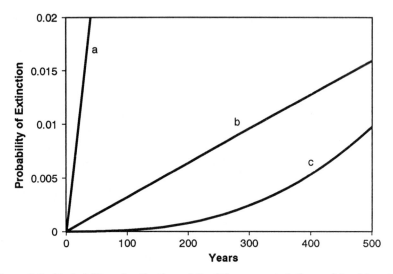

Figure 5.5   Probability of extinction of the Virunga populations of the Mountain Gorilla, assuming no dispersal between populations and therefore no recolonization of populations that become extinct: curve *a*, 1 population, $\frac{1}{3}$ size; curve *b*, 1 population, full size; curve *c*, 3 population, each $\frac{1}{3}$ size, with 0 correlation of the environment between them.

However, when the question is how to choose between one large or several smaller reserves, the answer is not straightforward. The answer to such a question depends on several interdependent factors including:

1. the difference between the extinction probabilities of a small and a large population;
2. the number of populations;
3. the dependence, or correlation between the environmental conditions for different populations; and
4. the probability of recolonization of an extinct population with individuals from other populations.

As expected, species extinction probability decreases with increasing numbers of populations and with decreasing dependence (correlation) among them. Occupancy models provide a quantitative method for incorporating these factors for estimating the relative risks of one large and several small populations.

Figure 5.6 shows the effect of different levels of correlation between the populations for the 3-population system. In this figure, recolonization probabilities are zero, and a 1-population system (one large population, three times the size of a single population) is included for comparison.

Similarly, Figure 5.7 shows the effect of recolonization. In this figure, the environmental fluctuations are uncorrelated ($r = 0$), and all six

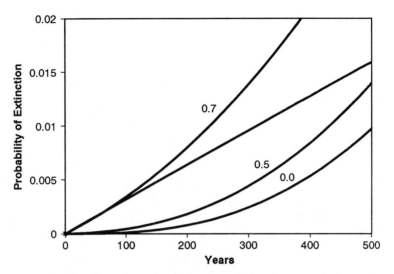

Figure 5.6 Probability of species extinction of the three-population system at different levels of environmental correlation from 0 to 0.7, again assuming no recolonization. The curve for the case of no correlation of environmental conditions between patches ($r = 0$) is curve *c* in Figure 5.5. The straight line represents the single large population (curve *b* in Figure 5.5).

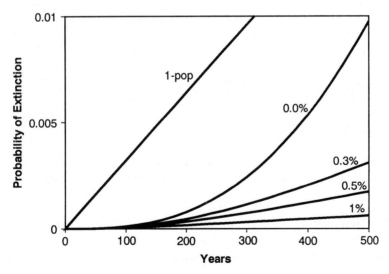

Figure 5.7 Probability of species extinction of the three-population system at different levels of recolonization probability from 0 to 1%. The recolonization rates between all population are equal in each case, and environmental correlation is assumed to be 0.

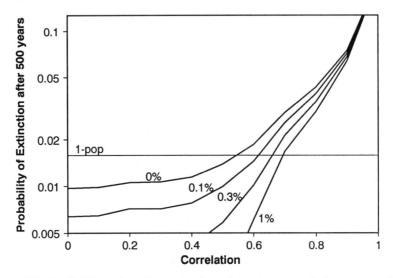

Figure 5.8   Probability of species extinction of the three-population system after 500 years at different levels of recolonization probability (from 0 to 1%) and environmental correlation (from 0 to 1). The straight line represents the single large population (curve *b* in Figure 5.5).

recolonization probabilities among three populations are assumed to be equal.

So far, the two factors that underly the occupancy model for the Mountain Gorilla – the chances of recolonization and the correlation between environmental conditions – have been treated separately. Figure 5.8 shows the combined effects of recolonization and environmental correlation for the 3-population system. In this figure, the probability of species extinction after 500 years is given as a function of 11 levels of correlation from 0 to 1 and four levels of recolonization probability from 0 to 1%. The horizontal line shows the extinction probability of a 1-population system after 500 years (note that the vertical scale is logarithmic). An important point to note is that when environments are highly correlated ($r > 0.9$), the extinction probability is very similar for all levels of recolonization (Figure 5.8). When environments are perfectly correlated ($r = 1$), the extinction probability is the same for all levels of recolonization.

The interaction between dispersal and correlated environments is interesting, not least because the question of whether a single large population is preferable to several small ones may depend on the time frame within which the question is posed. A high correlation among the environments experienced by three small populations of Mountain Gorillas makes them more vulnerable to extinction than the single large population. For example, if there is no dispersal among the populations, the 3-population system in highly correlated environments ($r > 0.7$) have no advantage over the 1-population system (Figure 5.5). However, with lower correlations, the multiple population system initially has an advantage, which decreases with time; the curves for single and multiple population systems cross, so multiple populations eventually have a higher extinction risk. In Figure 5.6, a recolonization probability of about 0.3% seems to be enough to ensure a long-term ($\gg500$ years) advantage for the 3-population system over the 1-population system, given that the environments experienced by the three populations are uncorrelated.

Figure 5.8 summarizes the interactions between dispersal and environmental correlation for the Mountain Gorilla example. For the three small populations to have lower extinction risk than the single large population when there is no dispersal, the correlation must be less than or equal to 0.5; the correlation must be less than 0.6 if the chance of dispersal equals 0.1%, and the correlation must be less than 0.7 if the chance of dispersal equals 1%. Thus, we can define a minimum rate of recolonization necessary to make a multiple-population system less vulnerable to extinction than a single population of the same total size. The exact value for recolonization depends on the number of populations and the correlation between them.

5.2.7 Assumptions and limitations of occupancy models

There are several advantages to using the occupancy model of Akçakaya and Ginzburg (1991) over some of the earlier formulations.

1. The use of a specific correlation coefficient and a specific dispersal rate for each pair of populations introduces a more complete spatial structure.
2. The local populations are not assumed to be identical; the total extinction probability may be different for each population, summarizing the differences in their carrying capacities and population dynamics.
3. It is possible to compute the risk of extinction for any subset of populations, in addition to the risk of extinction of the entire metapopulation.

However, the model shares some of the limitations that are intrinsic to all occupancy models, such as characterization of each patch as either occupied or empty, thus ignoring the details of the events that occur within each population.

Empirical observations of metapopulations
Harrison (1991) reviewed the empirical literature on metapopulations and found few examples fit the conceptual framework of a set of populations persisting in a balance between local extinctions

and colonizations. Rather, she identified three broad categories of metapopulation dynamics:

1. mainland island populations in which persistence depends on the existence of an extinction-resistant population;
2. patchy populations in which dispersal is so high that the system is effectively a single extinction-resistant population; and
3. non-equilibrium metapopulations in which local extinction occurs in the course of a species' overall decline.

Her observations suggest that information on both local and regional dynamics are essential, together with dynamic models with the characteristics of the occupancy models described above. This level of detail will require a great deal of data.

The most important limitation of all the metapopulation models discussed in this chapter pertains to parameter estimation. Although parameter estimation is the most crucial and limiting step for any population model, it is especially difficult for probabilistic models. Measuring probabilities of extinction is difficult, even for single populations. As a result, models that use an extinction probability as a parameter are generally difficult to apply to real cases. One way to overcome this problem is to estimate extinction probabilities of individual populations from another model that uses life history data as parameters and extinction probabilities are output. This is the approach we used in developing the model for the Mountain Gorilla.

When we looked at the effect of correlated environments, all we did was apply a range of correlation values among local extinction probabilities per unit time. This approach amounts to an analysis of the sensitivity of the total (species) extinction probability to the degree of correlation. It would be much more difficult to measure the actual degree of correlation in the Mountain Gorilla metapopulation.

One important complication is that there may be a difference between the correlation among **extinction probabilities** and the correlation we observe among actual **extinction events** (Gilpin, 1990). Occupancy models generally use correlations among extinction probabilities, and they are very difficult to estimate. Extinction events, on the other hand, can in principle be estimated from field observations, but even this would require the observation of many extinction events, hardly a practical exercise for most rare or threatened species.

The analysis above illustrates some of the trade-offs with parameter estimation that arise when we introduce greater realism in models. Although the model used in this analysis has some advantages over the other generalizations of the Levins' model, it has one important drawback: the number of parameters. As the number of patches increases, the parameter estimation process becomes too complicated for practical purposes. For just seven populations (the number of populations of *Banksia cuneata* for which we build a model in Chapter 6) there are 2^7 (128) possibles states for the metapopulation, as well as 42 ($7^2 - 7$) dispersal probabilities and 21 (($7^2 - 7$)/2) correlations between local extinction probabilities. On the other hand, for metapopulations composed of a small number of local populations the model provides a fast and relatively realistic method for estimating species extinction probabilities. The number that can be handled in an occupancy model depends on the amount and quality of available data, but things become difficult for more than about five populations.

For larger metapopulations a more convenient method is a simulation approach based on local population dynamics. This approach will be described in the next section.

5.3 A POPULATION DYNAMIC MODEL

Populations can be characterized by more than just their presence or absence in a patch. Populations differ in size, growth rate, density dependence, age or stage structure. There are several approaches to modelling dynamics of single populations that can form the basis of a metapopulation model to address the questions of predicting extinction risks and designing nature reserves. One approach is based on the generalized birth and death process (Kendall, 1948). The models based on birth and death processes mostly concentrate on demographic stochasticity, and predict the expected time to extinction using population (rather than per capita) rates of birth and death (MacArthur and Wilson, 1967; Richter-Dyn and Goel, 1972; Leigh, 1981; Goodman, 1987a). We discussed this approach briefly in Chapter 2.

Quinn and Hastings (1987) derived a formula based on the single-population model developed by Leigh (1981) and suggested that the best way to divide the reserves is to have $m = \sqrt{N}$ reserves, where N is the total number of individuals. Gilpin (1988) pointed out that their formulation does not include the consideration of environmental correlations among localities and dispersal and recolonization between populations. The question of whether a single large or several small reserves offer more protection for species can better be answered by taking these two important factors into account.

There are several other types of single-population models that use different types of species-specific information, such as density dependence and demographic and environmental stochasticity (Lewontin and Cohen, 1969; Levins, 1969; May, 1973b; Roughgarden, 1975; Boyce, 1977; Tuljapurkar and Orzack, 1980; Ginzburg *et al.*, 1982; Lande and Orzack, 1988). When additional factors such as age structure and correlated variation of survival and fecundity between age classes are also incorporated into stochastic population models, analytical solutions are not possible and simulations must be used (Shaffer, 1983; Shaffer and Samson, 1985; Ginzburg *et al.*, 1984, 1990; Ferson *et al.*, 1989). This approach, described in detail in previous chapters, can be extended to the study of metapopulation dynamics by a simulation model with several populations, and with correlated environmental variation and dispersal between populations. Such a model will give the species extinction probability (i.e. the probability that all populations are extinct) under different conditions of environmental correlation and recolonization among populations.

Below we shall describe a metapopulation model which incorporates species- and population-specific factors into the prediction of species extinction risks. The aim is to translate information about the biology of a species and the particularities of its distribution into estimations of extinction risks that would be useful for reserve design.

5.3.1   Variables and parameters of the model

The system of multiple populations in this model is represented by a number of single species population models that describe the local population dynamics at each location at whatever complexity and level of detail is appropriate for each population. For instance, the specific population dynamic model at each location can be density-independent Malthusian (exponential) growth, or density-dependent growth according to some function such as the logistic or Ricker function. It may also include Allee effects, and both demographic and environmental stochasticity.

For example, we may decide that the most appropriate model to describe the change in size of a particular population is the model for exponential population growth. For density-independent growth with environmental stochasticity, the growth of the ith population in an n-population system is simulated by the difference equation

$$N_i(t + 1) = N_i(t)\lambda_i - \sum_{j=1}^{n} E_{ij} + \sum_{j=1}^{n} E_{ji} \qquad (5.9)$$

where n is the number of populations, N_i^t is the number of individuals in population i at time t, λ_i^t is the growth rate of population i at time t, and E_{ij} is the number of individuals emigrating from population i to population j. The growth rates λ_i^t of the n populations are random variables with means $\bar{\lambda}_i^t$ and standard deviations σ_i, and are correlated according to an $n \times n$ correlation matrix **P**. Thus the elements of the correlation matrix, ρ_{ij} give the correlation between the growth rates of the ith and jth populations through time. If the difference equation model

$$N_i(t + 1) = N_i(t)\lambda_i \qquad (5.10)$$

is viewed as a discrete time approximation for the differential equation

$$\frac{dN_i}{dt} = r_i N_i \qquad (5.11)$$

where r_i is a normally distributed random variable, then λ_i should be lognormally distributed. This means that when you use Equation (5.10) to model the dynamics of a population (as we did in the model of the rhinoceros population in Chapter 2), the random variation in the growth rate should be chosen from a lognormal distribution (see Appendix, Section A.2.4). In the deterministic case, $\lambda_i = 1$ describes a stationary population.

The local population dynamic models can include more detail, such as demographic stochasticity, stage structure and density dependence. The important point is that populations are first modelled separately and are then linked with the incorporation of two spatial factors, correlated environmental fluctuations and dispersals.

5.3.2 Correlations

One way to model the effect of correlated environmental fluctuations is to describe the dynamics of each population with functions of environmental variables such as temperature, precipitation, etc. For example, the growth and reproduction of giant kelp may be described with equations for water temperature, light availability and nutrient concentrations (Burgman and Gerard, 1990). Given records of these environmental variables for each patch, it would be possible to calculate the correlations between patches. This correlation structure would then translate into a correlation structure for the population dynamics through the functions. However, in most cases it is not possible to model the dynamics of a population using precise functions of environmental variables, usually because of a lack of data.

An alternative approach is to use the observed (or estimated) correlation among demographic variables for different populations. The simplest case is Equation (5.9), where there is only one demographic variable λ_i for each population. In this case, the correlation structure is described by an $n \times n$ correlation matrix. To estimate the elements of this matrix, data on population growth rates for all populations over a number of years are needed. If such data are not available, but population growth rates are known to be affected by a certain environmental factor such as temperature or precipitation, then the environmental correlation among localities can be used to approximate the correlation among growth rates. This would assume that population growth at each location is perfectly correlated with the pertinent environmental variable. Alternatively, the product of the correlation coefficient of the environmental variable with population growth in each location, and the correlation of the variable among locations, will give the joint correlation needed for the $n \times n$ correlation matrix.

Sometimes data are available to estimate correlations among certain localities, but not among others. Assuming environmental correlations decrease with distance, then another approach can be used. In some cases, environmental correlation (and the correlation of environmentally induced fluctuations in the population growth rates) will be a monotonically decreasing function of the distance between the two localities. Available data can be used to estimate this functional form, or different functions can be tried to analyse the sensitivity of the results (the extinction probabilities) to the exact form of this function. Following this procedure, unknown correlations can be assigned values according to the distance between populations. This approach may be easy to implement, but it makes the assumption that correlations depend only on the distance between populations.

The correlation matrix is used in the model to generate correlated random variables. In the example given above, these variables are population growth rates λ_i with a specific mean and standard deviation for each population. A method for generating correlated random variables is discussed by Scheuer and Stoller (1962), and a method is given in the Appendix, Section A.3.2. Note that a correlation matrix must be symmetric, but not every symmetric matrix can be a correlation matrix. A correlation matrix must also be positive definite (see a textbook on linear algebra for a definition). If real coordinates for each population are used together with a monotonically declining correlation-distance function to estimate correlation coefficients, the resulting matrix will be a proper correlation matrix (in addition, the diagonal elements must be equal to 1).

5.3.3 Dispersal

Emigration from population i to population j is a function of dispersal rate, m_{ij}, from population i to population j, and the number of individuals in population i. In the case of density-independent emigration this function is linear, i.e. a constant proportion m_{ij} of individuals emigrate from population i to population j each year:

$$E_{ij} = m_{ij} N_i. \tag{5.12}$$

It can also be a non-linear function of the number of individuals, such as when the number emigrating or dispersing is zero under some threshold population size, or when the proportion of individuals emigrating increases at high population sizes. The effect of dispersal can then be described with an $n \times n$ matrix of dispersal rates (or functions). Unlike the correlation matrix discussed above, a dispersal matrix need not be symmetric, since the rate of dispersal may be different in two directions between the same two populations. This means that up to $n(n - 1)$ dispersal rates (or functions) must be estimated. This estimation procedure can be made simpler if it is known that the rate of dispersal is a function of the distance between two populations (see Section 5.1.2).

Below, we give an algorithm that represents the basic structure of the model. The details of its implementation will depend on the selection of functions for density dependence, dispersal, correlation and stochasticity. It is a simplified version of the algorithm used by Akçakaya and Ferson (1990).

Algorithm 5.1

1. Specify dispersal and correlation matrices; produce a variance–covariance matrix using the correlation matrix and the standard deviations of growth rates λ_i for each population, i.
2. For each replication, do steps 3 to 10.
3. Initialize population sizes.
4. For each time period, t, do steps 5 to 9.
5. Select the growth rates, $\lambda_i(t)$ using the variance–covariance matrix, the mean growth rates (and also the population sizes N_i if population growth is density dependent).
6. For each population i, predict the population growth, for example
$N_i(t) = \lambda_i(t)N_i(t - 1)$.
7. For each population i, predict the number of emigrants to each of the other populations, for example
$E_{ij} = m_{ij}N_i(t)$
m_{ij} may be a function of N_i if dispersal is density dependent.
8. For each population i, predict the number of individuals including population growth and dispersal
$N_i(t) = N_i(t) + \Sigma\, E_{ji} - \Sigma\, E_{ij}$.
9. Record the population size at t, the current time period.
10. Record minimum and maximum population sizes for this replication.
11. Calculate average population sizes (over all replications) at each time step; calculate probability of population decline and increase from the records of minimum and maximum population sizes.

5.3.4 A hypothetical example

To illustrate the potential uses of this population dynamics approach, we shall develop in this section a model for a hypothetical metapopulation consisting of seven populations. The spatial pattern of this metapopulation is shown in Figure 5.9, although any arrangement of groups is possible. The seven populations are located close to each other forming a compact system. There is dispersal (in both directions) between a population and its three nearest neighbours, except for the population in the middle, which exchanges migrants from all other six populations. In this example, we let each arrow represent the same dispersal rate. This rate is

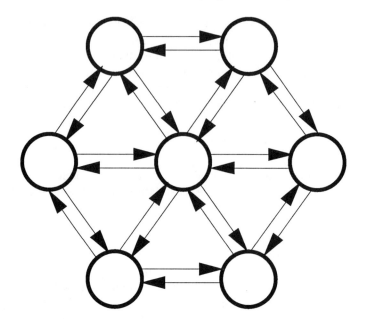

Figure 5.9   The spatial pattern of seven populations used in the simulations.

changed in different simulations to analyse the sensitivity of the extinction risk to the rate of dispersal.

The dispersal matrix below is for a specific case of 10% dispersal between any two connected populations.

$$\mathbf{M} = \begin{bmatrix} 0 & 0.1 & 0 & 0.1 & 0.1 & 0 & 0 \\ 0.1 & 0 & 0.1 & 0.1 & 0 & 0 & 0 \\ 0 & 0.1 & 0 & 0.1 & 0 & 0 & 0.1 \\ 0.1 & 0.1 & 0.1 & 0 & 0.1 & 0.1 & 0.1 \\ 0.1 & 0 & 0 & 0.1 & 0 & 0.1 & 0 \\ 0 & 0 & 0 & 0.1 & 0.1 & 0 & 0.1 \\ 0 & 0 & 0.1 & 0.1 & 0 & 0.1 & 0 \end{bmatrix}.$$

For this demonstration, we also analysed the sensitivity of extinction risk to the degree of correlation between the growth rate among populations. For this we used five functions of distance to calculate the correlation matrix (Figure 5.10). All these are negative exponential functions with a value of 1 at zero distance (see Equation (5.3) and Figure 5.4).

The species extinction probability in this model is expressed as a function of a threshold (i.e. some population size) below which the total size of the seven populations falls at least once during a 100-year period. The probability of falling below a threshold is calculated using the minimum total population sizes of each of the 1000 iterations.

Figure 5.11 shows the quasiextinction probability as a function of

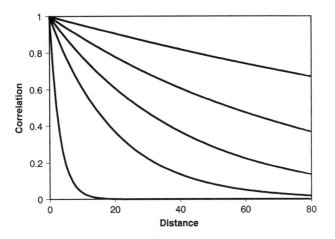

Figure 5.10 Degree of correlation among the growth rates of two populations are specified as a negative exponential function of the distance between them. Five different functions are used for different levels of correlation.

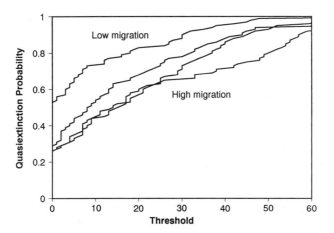

Figure 5.11 Quasiextinction risk curves for the metapopulation, given four different rates of dispersal ranging from 1 to 15% per year along each path in the diagram in Figure 5.9. The metapopulation has a significantly higher extinction risk with low dispersal. Extinction risk is less sensitive to changes in dispersal when it is at higher levels. Correlations between patches were fixed.

threshold for four levels of dispersal rate. As expected, a higher rate of dispersal leads to lower extinction risk since the extinct patches have a better chance of being recolonized. Another interesting pattern in this figure is that extinction risk is more sensitive to changes in dispersal rate at low levels of dispersal: the difference between the top two curves is much larger than the difference between the bottom two curves. This is

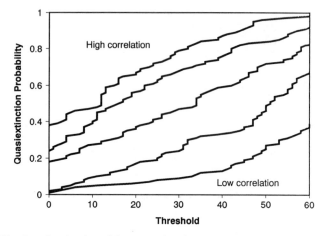

Figure 5.12   Quasiextinction risk curves for the metapopulation, for five different levels of correlation between environments. The metapopulation has significantly higher extinction risk when growth rates are highly correlated.

because at high levels of dispersal, the metapopulation is so well mixed that any additional increase in dispersal does not affect the extinction risk as much.

Figure 5.12 shows the extinction curves for five levels of correlation between growth rates of the populations. Each curve is produced with a simulation that uses one of the five correlation-distance functions in Figure 5.10. As expected, a low correlation leads to a low extinction risk. The reason, as explained in the first section, is that when the growth rates are uncorrelated, populations will increase or decrease independently of each other. When some of the populations decrease (or even become extinct) the others may not, and the small (or extinct) populations have a chance of being recolonized from the others. When the correlation is very high (the top two risk curves) the fluctuations are almost always in the same direction in all the seven populations, and the metapopulation 'behaves' like a single population: an extinct population cannot be recolonized because most (or all) of the others are also extinct at the same time period. This result shows that extinction risk is quite sensitive to the degree of spatial correlation in environmental fluctuations. Thus, ignoring correlations (i.e. assuming all populations grow independently), as most occupancy models do, will seriously underestimate extinction risks.

Figure 5.13 shows the effect of another assumption of occupancy models: equal carrying capacities (K) of all populations. In this figure, the two continuous lines were produced as in previous simulations, using a model of seven equally sized populations. The dashed curves were produced with a model of seven populations which have the same total

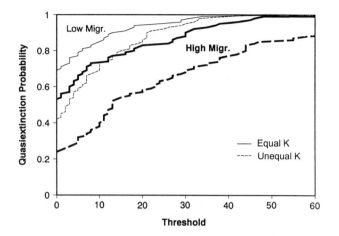

Figure 5.13 Quasiextinction risk curves for the metapopulation using populations of equal size (solid lines) and populations of unequal size (broken lines),
together with different dispersal rates. The metapopulation has a significantly
higher estinction risk when carrying capacities of local populations are not equal.
This effect is more pronounced when the dispersal rates are high.

(i.e. metapopulation) carrying capacity. However some of the populations in this (unequal K) simulation have larger and some smaller
carrying capacities compared to the populations in the other (equal K)
simulation.

The result shows that the metapopulation as a whole has a higher
extinction risk when the total carrying capacity is distributed unequally.
This is because of an asymmetry in the relative benefit of increasing K
versus the disadvantage of decreasing it. There is a disproportionate,
adverse effect of demographic stochasticity in small populations. This
result shows that the assumption that all populations are equal (as is
assumed in most occupancy models) may underestimate extinction risks.

Another interesting, and partly counter-intuitive, pattern in this figure
is that the effect of unequal K is more pronounced when dispersal rates
are high (bold curves). This may be because the small populations are so
extinction-prone because of high demographic stochasticity that they act
as population sinks. If this is the case, an increase in dispersal rate will
move individuals more quickly to these sinks. This may be the reason
that, at some threshold levels, the metapopulation with unequal K and
high dispersal has a higher risk than the one with equal K and low
dispersal.

This pattern points out an important complexity: the interaction
between factors, i.e. the determination of the effect of one factor by the
level of another. In this example, one factor (K) not only affects the
extinction risk, but the degree to which it affects the risks depends on

the level of a second factor, dispersal. Even this example showing the combined, interacting effect of only two factors is quite complicated. It would be difficult to anticipate this pattern without using a mathematical model. In addition, there are other factors (number of populations, their spatial configuration and the correlation of the environments they experience, to name a few) that we did not consider in combination with these two factors.

The problem gets even more complicated when one adds management options (such as re-introduction) or human impact (such as road building that increases fragmentation). It is impossible to make unconditional generalizations about whether it is better to have a single large or several small reserves; the decision depends on conditions such as environmental correlations and dispersal between populations. The conditions under which one management alternative would be preferable to another also depend on factors such as life history characteristics of the species, and how they are affected by environmental and demographic stochasticity. The model we described is designed to incorporate all these character- istics, and thus can be used in assessment of species extinction risks as a realistic tool.

The major simplifying assumption of this example is that all individuals in a population are similar. No age or stage structure is built into the model to describe differences in demographic properties of organisms as a function of their age or their physiological/developmental stage. The next level of complexity in models of this kind is represented by models incorporating such a structure within each population, in addition to the spatial structure that determines the interaction among populations.

Models have been developed that include age or stage structure in a metapopulation model (for example, Roff, 1974). Lefkovitch and Fahrig (1985) used a stochastic, age-structured model of a five-patch meta- population. They explored different topologies of connections between populations and demonstrated the importance of connectivity in meta- population survival. Fahrig and Merriam (1985) used a similar approach to model a metapopulation of white-footed mice (*Peromyscus leucopus*) in seven woodlots. Their model demonstrated that the spring growth of mice populations is at a higher rate in connected woodlots than in isolated woodlots – a finding that was corroborated by field data.

Although these models had age structure, they did not have a full spatial structure including correlations among populations. In addition, the random fluctuations in survivorship probabilities and fecundities were uncorrelated among age or stage classes. This may not be a realistic assumption for some species: when it is a 'bad' year for a specific popu- lation, survivorships (and perhaps fecundities) for all age classes in that population can be expected to be low. Furthermore, the assumption is not conservative. By modelling fluctuations of demographic parameters independently, these models underestimate the overall level of population fluctuations, because fluctuations in different age classes will tend to cancel each other. The result is an underestimation of extinction probabilities.

As demonstrated above, correlations among populations are expected to increase extinction risks of metapopulations. This is intuitive, and has been demonstrated by the use of both occupancy models (Harrison and Quinn, 1989; Akçakaya and Ginzburg, 1991) and unstructured population dynamic models of metapopulations (Akçakaya and Ferson, 1990). Thus, ignoring any of the two correlation structures will cause an underestimation of the extinction risks. Other factors important for population dynamics (such as demographic stochasticity and density dependence) have similarly been omitted from age- and stage-structured metapopulation models.

The next level of complexity in the spectrum of models of spatial population structure is a population dynamic model that includes stage structure, among-stage (within each population) correlations and among-population correlation structures. Clearly, metapopulation models at all levels of complexity will be useful in different cases. Their usefulness will depend on the biology of the species, the amount of available information on its dynamics and on the metapopulation structure, and on the questions that need to be answered.

One direction in metapopulation modelling may involve building explicit links between habitat models and metapopulation models that incorporate population dynamics. New developments in geographic information systems (GIS) and landscape ecology make it possible to predict changes in the spatial structure of a species' habitat. Such changes may be due to climatic changes, human or natural disturbances, or vegetation dynamics. These predictions may then be used to estimate future changes in the configuration, size and carrying capacity of habitat patches through time. This may then lead to the development of simulation models in which not only the dynamics of populations, but also the dynamics of the habitat patches they live in, are modelled.

5.4 SUMMARY

The environment does not affect the individuals in a population uniformly throughout the geographic range of a species. Whenever the environmental components that limit the distribution of a species are distributed patchily, or whenever the original continuous habitat is fragmented, the species is likely to be distributed in a number of populations, more or less isolated from one another, and probably experiencing some degree of different environmental conditions. To model the spatial component of the dynamics of populations, it is necessary to deal with the dynamics of interacting populations, or metapopulations.

There are two important aspects of metapopulation dynamics to consider: the correlation between environmentally induced fluctuations in the vital rates of different populations, and the rate of dispersal of individuals between populations. For many species, both of these factors may be modelled as functions of the geographic distance between populations. The chances of extinction of a metapopulation will depend on these factors, and on the interaction between them.

Occupancy models were developed from the work of Levins and they simplify the problem by treating patches as either occupied or unoccupied. This approach assumes that there is no local population dynamics, independent extinction events and equal rates of dispersal among patches. Even though these assumptions, on close examination, are tenuous and even contradictory, the model is a useful first approximation to metapopulation extinctions.

The model has been generalized by a number of people to account for correlated extinction probabilities (reflecting a degree of correlation between environments) and different population sizes. We present a recent development that includes both these components in a single occupancy model. The model uses local extinction probabilities and specific dispersal rates between pairs of patches as input, and employs a set of transition matrices to compute the chances of extinction of the metapopulation. It allows the evaluation of models with full spatial structure, although the number of parameters becomes daunting for even moderate numbers of populations.

Lastly, we introduce a model that includes the dynamics of local populations, as well as environmental correlations and dispersal. The model could form the basis of a new class of metapopulation models that account for the structure and correlations within populations. A hypothetical example serves to demonstrate that the best design of nature reserves for a species depends on the levels of correlation and dispersal between populations, as well as the number, size and dynamics of the populations. In general, high levels of environmental correlation and low levels of dispersal between populations will tend to reduce the chances of persistence of species, although the interaction of these factors with population size, and many other factors such as the configuration of patches, disease, or human intervention, may reverse these generalizations.

6 Conservation genetics

One of the goals of conservation is to predict and try to avert the genetic deterioration of species, to preserve species' potential for adaptation to both short- and long-term environmental variation, and thereby reduce their chances of extinction.

If, because of disease, loss of habitat or some other extrinsic factor, only two representatives of a species remain, common sense suggests there is trouble ahead. Quite apart from the vagaries of demographic accidents or further detrimental environmental variation, the species is at risk because of genetic uncertainty, especially if individuals normally avoid mating with relatives.

So far, we have assumed that populations are genetically homogeneous in space and time. In this chapter, we address the genetic composition of populations. We review the processes that lead to genetic deterioration and describe the information and equations needed to characterize genetic variability within and between populations, using an example of the rare Spanish herb, *Silene diclinis*. We provide the equations to estimate rates of loss of genetic diversity and effective population size necessary to avoid serious genetic deterioration, given different population structures in space and time. Lastly, we outline methods that combine the stochastic models for risk assessment developed in the previous five chapters with the ideas of population genetics. We apply these methods to a rare and restricted Australian shrub, *Banksia cuneata* and to populations of the white rhinoceros.

There are a number of excellent introductions to quantitative population genetics, two of which are by Maynard-Smith (1989) and Falconer (1989). We do not intend to reproduce all of the methods described in those texts. Rather, we outline the methods within the framework of risk assessment for conservation biology and use examples to show how they may be applied to predict the risks of genetic decline and its effect on population viability, and point to developments of quantitative genetic methods suitable for more complex applications.

6.1 CONSEQUENCES OF THE LOSS OF GENETIC DIVERSITY

Fisher's (1930) fundamental theorem of natural selection tell us: 'The rate of increase in fitness of any organism at any time is equal to its genetic variance in fitness at that time.' Genetic variation is the raw material on which natural selection acts. When genetic variation is lost from a population, its rate of increase in fitness and its ability to adapt to environmental change are reduced. Such losses may have important consequences for the risks of population decline.

6.1.1 Adaptation to change

Two individuals will not carry all of the genetic variation present in their ancestors. They will be fixed for some alleles that were polymorphic in the parent population with the result that variability will have been lost. For example, if the two are homozygous for loci that determine size, photosynthetic pigment, or resistance to cold, if these characters are

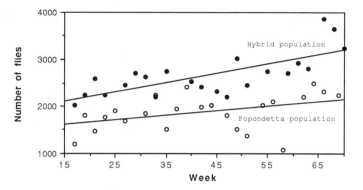

Figure 6.1   Population size of two populations of *Drosophila serrata*. The hybrid population is a hybrid between the Popondetta strain and a Sydney strain. (After Ayala, 1968.)

genetically based and heritable, and if they carry the same alleles then the variation in these characters will be lost from the species. Populations that are reduced to very small sizes may adapt less readily to changes in environmental conditions and may be less tolerant to extremes of environmental variation.

Thus, the preservation of genetic diversity is important because diversity enhances the ability of a species to adapt and is essential for long-term evolutionary potential (Fisher, 1930; Wright, 1948). There is experimental evidence to support the notion that genetic variation *per se* enhances the chances of persistence of populations through the ability of a population to adapt to environmental change. Ayala (1968) used population size in an artificial laboratory environment as a measure of adaptation of *Drosophila* populations to novel environmental conditions, and compared a hybrid population with one of its parental strains. The populations were at the carrying capacity of their artificial environment and larger population sizes simply greater efficiency in transforming limited food and space into biomass. Hybrid populations between different strains of a species, if based on many parents, will contain most of the genetic variation present in both parental strains. The hybrid population in Ayala's study performed consistently better (Figure 6.1).

In the short term, pre-adapted individuals (individuals that, by chance, have characteristics that predispose them to relatively high fitness in the face of environmental change) are more likely to be present in a variable population, increasing the chances of surviving rapid change. For example, the productivity and size of experimental populations is higher in *Drosophila robusta* and *D. serrata* collected near the centre of their geographic ranges, in locations where their genetic variability is greater (Ayala, 1968).

Long-term environmental change will elicit an adaptive response in a

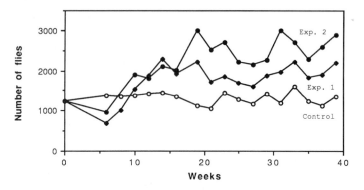

Figure 6.2 The effect of radiation on the growth of three populations of *Drosophila serrata*. Both experimental populations were irradiated whereas the control population was not. (After Ayala, 1968.)

species through natural selection only if there is heritable genetic variation among individuals. Populations with greater initial variability perform better and have a higher average growth rate in the long term because natural selection is more efficient in populations where more genotypes are available for selection. In another experiment, two populations of *Drosophila serrata* were irradiated and their populations monitored by Ayala (1968). The radiated populations declined in numbers over the first few weeks, presumably because of the elimination of individuals carrying deleterious mutations (Figure 6.2). Subsequently, the two experimental populations grew considerably faster and attained larger average population sizes than the control which was not irradiated. Natural selection resulted in better adapted genotypes in the irradiated populations.

6.1.2 Inbreeding depression

When populations are small, or when there is a propensity for individuals to mate with close relatives, offspring have a chance of acquiring alleles at a locus that are identical because they are derived from a common ancestor, transmitted through each parent. One of the reasons why loss of heterozygosity is important for the conservation of species is that all populations carry recessive alleles and a proportion of these alleles are lethal when homozygous and a larger proportion are deleterious. Slatis (1960) found an average of six lethal recessive genes were carried by each individual in a population of European bison and Templeton and Read (1983, 1984) found an average of five lethals in individuals of Speke's gazelle (*Gazella spekei*). James (1982) found in *Isotoma petraea*, a small perennial shrub restricted to rocky substrate in Western Australia, that 90% of the ovules fail to develop into seeds because of the expression of recessive lethals. The increased homozygosity from breeding between relatives usually results in reduced fertility, survivorship, disease resistance and growth rates, and is termed inbreeding depression (see, for example, Ralls *et al.*, 1986; Foose and Ballou, 1988).

We touched on the relationship between inbreeding and fitness in Chapter 1, in the section titled 'The genetics of species at risk'. We noted that inbreeding depression, expressed as reduced survivorship and fecundity, had been detected in numerous species of plants, insects, and mammals including cats, ungulates and primates (Slatis, 1960; Ralls and Ballou, 1982a,b, 1983; Ballous and Ralls, 1982; Templeton and Read, 1984).

Genetic variation in cheetah populations
O'Brien (1989) and O'Brien *et al.* (1986) studied the genetics of two remaining populations of the cheetah (*Acinonyx jubatus*) in Africa. The southern population of about 1500 individuals is almost devoid of genetic variation. Male fertility, juvenile mortality, and susceptibility to disease are all adversely changed in this population. O'Brien *et al.* conclude that the lack of variation in the population is associated with detrimental effects on traits, which results in lower chances of persistence of the population.

In general, if close relatives of habitually outbreeding species are mated, there is a tendency for their offspring to be less virile and less viable than the offspring born to unrelated parents. This has potentially important consequences for small populations. If two survivors remain in a species, the first generation will not be inbred, but the second will be the result of mating among very closely related individuals. Genetic variability plays a direct role in determining probabilities of survival and reproduction and thereby affects the chances of persistence of populations. Loss of genetic diversity may have short-term effects on population dynamics by reducing traits closely associated with fitness, such as the number of viable offspring produced per adult.

An inbreeding coefficient is a measure of the number of genes in an individual that are identical because they came from the same ancestor. Effectively, it reflects how closely the average parents in a population are related to one another. Figure 6.3 shows how the fertilities of three very different species, namely mice, maize and fruit files, are affected when they become more inbred. Falconer (1989, Table 14.1) and Ralls and Ballou (1982a,b; 1983) list numerous other examples of reduced growth rate, development, and fecundity that result from inbreeding.

The generality of this relationship allows us to use it in models of the dynamics of populations. There is a well-developed literature that de-

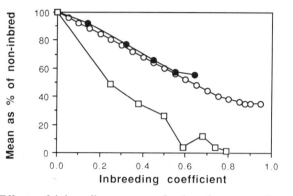

Figure 6.3 Effects of inbreeding on reproductive characters. Open circles are yield of seed in maize (*Zea mays*) after consecutive selfing. Closed circles are litter size in mice: mean number born alive in first litters, double first cousin and full sib mating. Squares are fertility in *Drosophila subobscura*: mean number of adult progeny per adult per day, consecutive full sib matings (after Falconer, 1989, Figures 14.1 and 14.2). The slope of the curves for maize and mice is about −0.8, and for *Drosophila* it is about −1.2.

scribes the calculation of the degree of inbreeding in a population. The equations make assumptions that mean the inbreeding can be estimated based on information about such things as the population size, the sex ratio and the basic reproductive behaviour of the members of the population (i.e. is mating based on complex social interactions; is distance between individuals important; or is mating more or less random?). Below, we shall explore these methods and apply them in a model of the dynamics of a population.

The relationship between effective population size and loss of genetic variability is relatively well understood but despite a number of empirical studies, the relationship between genetic variability and population viability is less well understood. Empirical evidence suggests a close relationship between the degree of inbreeding in a population, and measures of viability and virility. Soulé (1981) and Allendorf and Leary (1986) reviewed studies that link survival, disease resistance, growth, development rate and developmental stability to heterozygosity at a single or a few loci and found most studies demonstrate or suggest plausible mechanisms for the effect of the gene on fitness. When considering total reproductive performance, a decline in fitness of about 25%, for example, may result from a loss of 10% of the heterozygosity in diploid, outbreeding species that have not previously been inbred (Frankel and Soulé, 1981).

Inbreeding depression may be caused by the increased chances of expression of detrimental recessive alleles, by reduced levels of fitness in homozygotes, or by a reduction in variability of offspring with the result that the chance of an individual's progeny surviving sudden environmental change may be reduced. There is little doubt that heterozygosity *per se* has selective value. Many unusual breeding systems and cytogenetic attributes may be directly attributable to selection for mechanisms that conserve genetic variation in the face of enforced inbreeding in small populations (see Darlington, 1958; James, 1982). While inbreeding depression does not always result from inbreeding (Templeton and Read, 1984), increased homozygosity and the tendency of deleterious alleles to be recessive are the genetic basis for the loss of fertility and viability that almost always results from inbreeding in domestic and natural populations (Falconer, 1989, p. 25).

6.1.3 Outbreeding depression

Shields (1982) described another genetic mechanism that may lead to reduced fitness. If species in nature are subdivided into small, inbred local populations and if there is little migration between these populations, then different populations may have different coadapted gene complexes. The set of genes unique to each population will confer relatively high fitness as long as it remains intact. If mating occurs between individuals with different coadapted gene complexes, the locally adapted complexes break down and lowered fitness results, a phenomenon Shields termed *outbreeding depression* (Shields, 1982; Templeton, 1986).

Outbreeding depression is caused by adaptation of geographic populations to local environmental conditions, and by intrinsic coadaptation, in which genes in a local population adapt to the genetic environment defined by other genes. James (1982) suggested that local gene pools in isolated populations must be coadapted for their lethal recessive alleles. Different gene pools may be coadapted differently, and hybridization between different local populations may expose the recessive lethals, thereby reducing the fitness of the populations.

6.2 DRIFT, RISK AND GENETIC DIVERSITY

Genetic drift is a sampling phenomenon. In a real population, a finite number of alleles each generation are drawn from the parental gene pool. Recombination during meiosis is usually random and, in the absence of selection, a random sample of the available genes is passed to the next generation.

Gene frequency will drift in much the same way as population size in the random walk described in Chapter 1. Note the similarity between the random walk in Figure 1.1, and the random fluctuations in gene frequency illustrated in Figure 6.4. The amount of variation in gene frequencies in a population from one generation to the next is directly related to population size. The smaller the population, the larger the

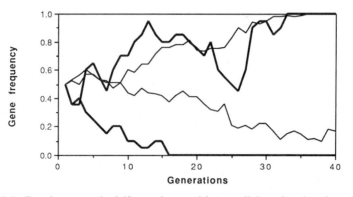

Figure 6.4 Random genetic drift at a locus with two alleles, showing the relative frequency of one of those alleles. Two replications of a sexual population with 5 females and 5 males are represented by the heavy lines. Two replications of a sexual population with 25 males and 25 females are represented by the thin lines. Each replication was begun with the two alleles distributed equally among males and females, each with a frequency of 0.5.

variance. We can see this in Figure 6.4 where gene frequency in the population of ten individuals appears to be more variable than gene frequency in the population of 50 individuals.

The simplest conceptual model for population genetics is one in which there is no selection and there are no complicating chromosomal mechanisms. The allelic frequencies 0 and 1 are absorbing barriers and genetic drift will act so that, eventually, all loci will become fixed for one or other of the alleles and the mean level of heterozygosity in the population will fall. Genetic drift is a fundamental property of populations. The smaller the population, the more rapidly gene frequencies will drift and the greater the chances they will reach one or other of the absorbing boundaries within a given period of time.

Sooner or later, random drift will fix an allele at any polymorphic locus. The variance in gene frequency after t generations is

$$s_q^2 = p_0 q_0 \left[1 - \left(1 - \frac{1}{2N} \right)^t \right] \tag{6.1}$$

where N is the number of breeding adults in a population that mates at random (more strictly, it is the effective population size, a concept we treat in more detail below). The variance calculated in this equation represents the uncertainty in predicting the frequency of an allele. The expected frequencies of the three phenotypes of a two-allele locus in a population are given by (Falconer, 1989)

$$AA = p_0^2 + s_q^2$$
$$Aa = 2p_0 q_0 - 2s_q^2$$
$$aa = q_0^2 + s_q^2.$$

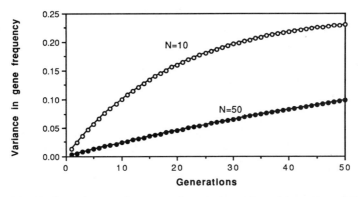

Figure 6.5 Variance in gene frequency for a locus undergoing random drift. The open circles show variance at the locus for a population of 10 individuals and the closed circles show variance at the locus for a population of 50 individuals. Both curves were generated using Equation (6.1).

If a population is very small, there will be fewer heterozygotes in the population than if the population is larger, because the magnitude of the variance in gene frequencies is related inversely to the size of the population (Equation (6.1)). Genetic substructures within a population will result in fewer heterozygotes than expected if the genetic variation in the population as a whole is sampled. Thus, sampling variance causes an excess of homozygotes in a species as a whole, even if Hardy–Weinberg expectations are met within subpopulations. The increase in total genetic variance due to the fixation of alleles in different populations within a species is known as the Wahlund effect (see Futuyma, 1986, and Falconer, 1989).

It is possible that a population of two individuals does not lose any genetic variation present in a previously large parent population, because none of the alleles at heterozygous loci happen to wander to fixation (Figure 6.4). The loss of an allele from a population is a chance event and the risk is larger in a small population. The risk depends not only on the size of the population, but also on the time frame we specify for the event to occur. None of the alleles in Figure 6.4 was fixed after five generations, but three alleles were fixed after 40 generations.

Consider a locus like the one in Figure 6.4. There are two alleles present at one locus in a population, and each allele has a frequency of 0.5 in the first generation. Note that the variance increases with time, so that our predictions become less certain the further into the future they are made. The allele frequencies in each generation are a random sample of the frequencies in the parent generation. If the genes drift randomly for 50 generations, and the process is repeated a large number of times, always starting with the frequency set at 0.5, Equation (6.1) gives the expected variance in gene frequency after each generation.

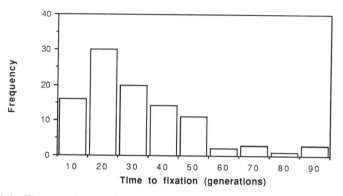

Figure 6.6 Times to fixation for an allele in a population of 10 individuals, based on 100 replications. The class labels are the upper class limits. Thus, in 16 out of 100 replications, the locus was fixed for one or other of the alleles within 10 generations.

The elevated variability of the allele frequencies in the smaller population is reflected in the greater variance for that population (Figure 6.5). In both cases, the mean (expected) gene frequency remains at 0.5. However, if a population is smaller, we can be less certain of predicting the relative frequency of an allele in the future.

Using the model for genetic variation in a population that gave Figure 6.4, we repeated the simulations of genetic drift many times, and recorded the time at which alleles were fixed, in much the same way as we recorded the time to extinction in the demographic models. These data were used to produce a histogram of times to fixation (Figure 6.6) for an allele that drifts randomly, and commences with a frequency of 0.5.

The distribution of time to fixation for a locus is skewed to the right (Figure 6.6), just like the times of extinction for a population in a randomly fluctuating environment. The mean time to fixation in the example above was 27.4 generations. However, this statistic will be misleading if you are interested in predicting when a population will become homozygous for a locus. A total of 58% of the replications were fixed for an allele before 27 generations, so it is likely that the population will be fixed for an allele before 27 generations have been completed.

To emphasize that genetic change is a stochastic process analogous to demographic change, we can produce risk curves that show the chances that an allele will be lost or will become rare in a population. We recorded the smallest frequency of one of the two alleles during each replication of the simulation for gene frequency drift. We used these data to construct risk curves for gene frequencies under three different scenarios. The first was a population of ten individuals, and gene frequencies were generated over a period of 100 generations. In the second scenario, the population size was increased to 50 individuals and, as a result, the

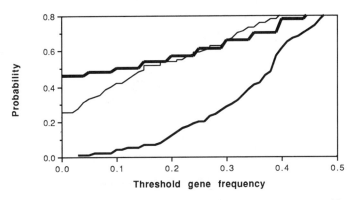

Figure 6.7 Risk curves for genetic drift at a single locus, based on 100 replications of the model. The relative frequency of the two alleles at the locus in the
first generation was 0.5. The heavy line represents the chances that the allele
drifts to a frequency as low as the threshold at least once in 100 generations in a
population of 10 individuals. The thin line represents the chances of drift in a
population of 50 individuals. The medium line represents the chances that the
allele drifts to a frequency as low as the threshold at least once in 20 generations in
a population of 50 individuals.

chance that the locus becomes fixed for the allele within 100 generations
is decreased from just under 0.5 to about 0.25 (Figure 6.7). In the third
scenario, the population size was left at 50, but the time horizon was
reduced to 20 generations. Reducing the time horizon to 20 generations
reduced the chance of fixation to nearly zero.

The discussion above deals with the loss of single locus genetic variability. Quantitative characters differ from single locus traits in a number
of important ways. Mutation rates are much higher. Selection will probably change the predicted effects of small population size on single locus
variation, but will have limited effect on quantitative characters (Lande
and Barrowclough, 1987). The response of a quantitative character to
selection will depend on the amount of heritable variation. Inbreeding
reduces single locus variability and it reduces the heritable genetic variation in quantitative traits (Figure 6.8). If natural selection acts on the
trait, the reduction in heritability will be less than expected.

Founder events and genetic bottlenecks are variations on the theme of
inbreeding and genetic drift. They are identical in the effect they have
on genetic variance in a population. In small populations, the loss of
heterozygosity occurs at a rate of approximately $1/(2N)$ per generation.
In a number of rare and restricted species, notably mammals, genetic
variability is low. In some of these cases, it has been possible to point to
severe reductions in population size in the past as a cause of loss of
variability. Bonnell and Selander (1974) use a population bottleneck to
explain the genetic status of elephant seals and O'Brien *et al.* (1986;

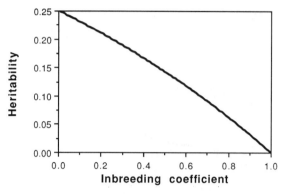

Figure 6.8 Reduction in heritable quantitative variation as a function of the inbreeding coefficient. The curve is generated by the equation $h_t^2 = h_o^2(1 - F_t)/(1 - h_o^2 F_t)$ where h_t^2 and F_t are the heritability within populations and the inbreeding coefficients at time t, and where h_o^2, the original heritability of a trait in the base population, is 0.5 (Falconer, 1989, p. 267). The formula assumes no non-additive variance and no selection.

O'Brien, 1989; cf. Pimm *et al.*, 1989) use it to explain the lack of variability in a population of cheetahs. Falconer (1989, pp. 54–60) and Maynard-Smith (1989, pp. 93–123, 144–61) provide more detail on the effects of genetic drift on the genetic diversity of populations.

Conversely, genetic variability in some rare and restricted species is relatively high (Moran and Hopper, 1987). If a population crashes, and quickly recovers, only a small proportion of the heterozygosity present before the crash will be lost. In the same way, founders of new populations in vacant habitat will not suffer great loss of heterozygosity if the population size increases rapidly. More importantly, the real world is not one of random mating, random recombination during meiosis, and sampling error imposed on a limited number of alleles from generation to generation. The reproductive strategies, behavioural constraints and chromosomal mechanisms of species can hinder or overcome the rate of loss of genetic variability. For example, James (1982) argues that there is intense selection for the evolution of chromosomal mechanisms that enhance genetic variability in populations of the plant *Isotoma petraea*. These mechanisms are usually apparent in species where there is self-compatibility or where the species is adapted to living in small populations.

The terms and concepts we defined in Chapter 1 to deal with demographic and environmental uncertainty serve equally well to describe uncertainty in the genetic make-up of populations. Soulé *et al.* (1986) recommend conservation and captive breeding programmes should aim to preserve 90% of the average heterozygosity of the initial population for the next 200 years. This statement is made in the same spirit as the specification of acceptable risks for demographic studies. It provides a benchmark against which to evaluate results and we may proceed to estimate the population size necessary to ensure the preservation of the genetic variability.

The difficulty in applying this concept to natural populations is that variability is partitioned among different populations. One way to maximize the variability in a species is to allow different populations to become fixed for different alleles. There will be no variability within populations but the variability between populations will more than compensate (Maruyama, 1972). However, the chances are reduced that the species will cope with long-term environmental change because there will be no variability within any of the populations on which natural selection can act. Thus, maximization of variability for its own sake will rarely be a useful approach.

A further difficulty is that the conservation of genetic variation involves both the preservation of heterozygosity and the preservation of alleles. These objectives may conflict. Alleles are best preserved in subdivided populations with no migration between them. Heterozygotes are best preserved in the short term in single populations, and in the long term in subdivided populations with high rates of migration between them (Boecklen, 1986). Transient reductions in population size may result in substantial loss of alleles, especially rare ones, whereas the loss of heterozygosity would not be severe (Frankel and Soulé, 1981). The preservation of alleles may be more important because heterozygosity can be increased by an increase in gene flow between populations, whereas lost alleles cannot be regained.

There is no simple procedure for applying risk assessment to questions of genetic variability, in contrast with the conceptual ease with which we can apply risk assessment to the demography of populations. When the focus is on the dynamics of the population, it is better to have more individuals in a population than fewer. When the focus is on genetic variation, more alleles may be better than fewer alleles, but it matters whether variation is between populations or within populations. There is no general approach to this problem that will satisfy all cases. Furthermore, there is a danger in relying on measures of genetic variation as indices of genetic status. All alleles are important in the sense that any one of them could provide the phenotype that will resist some future environmental perturbation. The most desirable distribution of genetic variation will depend on the particular spatial structure and reproductive behaviour of a species. The index used to represent genetic variation will depend on what aspect of genetic variability you intend to conserve.

6.2.1 Sampling genetic diversity

Often it is important to know how genetic diversity is distributed within a species. This kind of information has particular relevance when we are faced with choosing to preserve just a few populations of a patchily distributed organism, or when we consider moving individuals between patches. The measurement and interpretation of genetic variability within natural populations provides the information necessary to predict the impact of such management activities on the genetic resources of a species.

The vast majority of data on the genetic structure of populations comes from isozyme studies. Isozymes are soluble enzymes, the product of structural genes, and they are assumed to be coded exclusively by single loci. Shaw and Prasad (1970) and Lewontin (1974) review the methods and interpretation on isozyme data. Isozyme studies explore the variability of single locus traits, and are usually assumed to be a means of direct evaluation of genetic variation. This is despite the fact that most major phenotypic changes in population evolution are the result of the accumulation of quantitative polygenic changes in existing phenotypes. DNA technology includes mitochondrial and nuclear ribosomal DNA restriction site mapping and related techniques and it is an alternative to isozyme techniques. It provides valuable tools for measuring genetic variation within and between natural populations of plants and animals (Templeton et al., 1990; Schaal et al., 1991).

If recommendations are made for the genetic management of a species based on a survey of 20–30 allozymes, there is an assumption that they provide a reasonable indication of the variability of the genome as a whole (Woodruff, 1989). There are few studies on the relationship between quantitative and allozyme variation but there appears to be no simple relationship between the two (Hamrick, 1983). A lack of allozyme variation does not mean that no variation exists (Lewontin, 1974; Hamrick, 1983; Templeton et al., 1990).

Another alternative is to use changes in the heritability of quantitative traits as an index of change in genetic diversity. Narrow sense heritability is the ratio of additive genetic variance to total phenotypic variance and there are numerous ways to estimate heritability of quantitative traits using the values of the traits in related individuals (see Falconer, 1989). Mutations of small effect in quantitative characters are more frequent by two or three orders of magnitude and are less likely to be deleterious than mutations of single genes of major effect (see Lewontin, 1974; Lande, 1981). The effect of drift on heritability and the variability of quantitative, multilocus traits is greater than the single locus case because there may be many loci, each drifting independently. Similarly, the rate at which heterozygosity is recovered by mutation is much faster for quantitative characters than it is for single locus characters (see Lande and Barrowclough, 1987, and references therein). The difficulty with using quantitative traits is that, while they are relatively easy to measure, it requires a great deal of work to determine their level of heritability.

Table 6.1   Allele frequencies for the esterase locus (EST-1) in populations of *S. diclinis*, after Prentice (1984). The labels fast, medium and slow are the relative speeds of anodal migration for the three alleles on an electrophoretic gel.

Subpopulation	Allele		
	Fast	*Medium*	*Slow*
A	0.12	0.72	0.16
C	0.04	0.75	0.21

6.2.2   Genetic diversity in *Silene diclinis*

Silene diclinis is a rare dioecious perennial herb, restricted to old terraced olive and carob groves near Jativa in the south-eastern part of Spain. There has been concern for the persistence of the species since at least the late 1800s and the population currently consists of about 600 individuals distributed among two main subpopulations. Prentice (1984) assayed the isozymes of the species and found important, unambiguous genetic variation at a locus that codes for the enzyme, esterase. Some of the results of her study are shown in Table 6.1.

The values in Table 6.1 are a sample of the genetic diversity of the species. The first step in evaluating the genetic status of a species is to understand the distribution of variation between and within populations. The total genetic variation within any species may be calculated (Nei, 1973, 1975) from

$$H_T = 1 - \sum_{i=1}^{k} \bar{x}_i^2 \qquad (6.2)$$

where H_T is total genetic diversity and $\bar{x}_i$ is the mean frequency of the ith allele at a polymorphic locus. In the *S. diclinis* population, the mean frequencies were 0.08, 0.735 and 0.185 for the fast, medium and slow alleles respectively, giving a value of 0.419 for total genetic diversity.

The total genetic diversity may be apportioned to variation within subpopulations and variation between subpopulations. H_s is mean genetic diversity within populations, and it is given by

$$H_S = 1 - \sum_{i=1}^{k} x_i^2 \qquad (6.3)$$

where x_i is the frequency of the ith allele within a population. H_S is the relative frequency of heterozygotes expected in a population if the assumptions of Hardy–Weinburg equilibrium were met within a population. We shall address the question of equilibrium assumptions of these genetic models below. More exactly, H_T is the probability that two

Table 6.2   Indices of genetic diversity for the EST-1
locus in populations of *S. diclinis* (after Prentice, 1984).

Index	Value
H_T	0.419
H_S (Subpop A)	0.442
H_S (Subpop C)	0.392
G_{ST}	0.005
Alleles/locus	1.3
Proportion of polymorphic loci	0.15
H_e	0.05

homologous genes from different subpopulations (demes) are identical
and H_S is the probability that two homologous genes from the same
subpopulation are identical (Nei, 1975). Between population differentiation, the genetic diversity between populations, is given by

$$G_{ST} = \frac{H_T - \bar{H}_S}{H_T}. \qquad (6.4)$$

The index G_{ST} is just the difference between total and within group
genetic diversity, expressed as a proportion of the total genetic diversity.
If genetic variation within populations is much the same as the total
genetic variation (H_T is approximately equal to H_S) then the differences
between subpopulations will be unimportant. Sometimes an alternative
measure of between population diversity is used,

$$D_{ST} = H_T - \bar{H}_S \qquad (6.5)$$

where $\bar{H}_S$ is within population genetic diversity averaged over all populations. When isozyme data are sampled from populations, the statistics
H_T, H_S and G_{ST} are often calculated using the above equations. Other
statistics of genetic diversity include the average number of polymorphic
loci, the average number of alleles per locus, and the allelic evenness
(H_e) for the species which is based on the means of the subpopulation or
population allele frequencies, averaged over all loci sampled including
those that are not polymorphic. The results of these calculations for the
subpopulations of *S. diclinis* are shown in Table 6.2.

One important thing to keep in mind when using these indices is that
the variances of H_S, H_T and G_{ST} are large unless calculated from a large
number of loci (Nei *et al.*, 1977). These indices are very poor indicators of
the behaviour of individual loci.

The small value of G_{ST} in Table 6.2 suggests that there is no important
genetic differentiation between subpopulations. This result is not surprising because subpopulations A and C are on the NE and NW faces of
the same hill, separated by no more than about 400 m (Prentice, 1984). It

Table 6.3   Measures of genetic differentiation within and among populations. All estimates are based on allozyme studies. Data here were summarized originally by Nei (1975) and Loveless and Hamrick (1984).

Taxa	No. of studies	H_T	H_S	G_{ST}
House mouse	1	0.097	0.086	0.119
Kangaroo rat	1	0.037	0.012	0.647
Drosophila equinoxialis	1	0.201	0.179	0.109
Horseshoe crab	1	0.066	0.061	0.072
Club moss	1	0.071	0.051	0.284
Autogamous plants	39	0.291	0.128	0.523
Mixed mating	48	0.242	0.174	0.243
Outcrossing plants	76	0.251	0.214	0.118

is very likely that reproduction occurs between at least some individuals from different subpopulations.

Prentice observed frequencies of heterozygotes of 0.276 and 0.286 respectively in subpopulations A and C. These values are much smaller than those expected under Hardy–Weinburg equilibrium, 0.442 and 0.392 respectively (Table 6.2). She explains the excess of homozygotes within the populations by the Wahlund effect, caused by the existence of still smaller (though unsampled) genetic neighbourhoods. She speculates that there may be restricted gene flow among these neighbourhoods allowing each to differentiate under microhabitat selection and/or genetic drift. The average proportion of polymorphic loci in outcrossing plants is about 0.4, compared to the value of 0.15 for *S. diclinis*. The low value of H_e for the total population and the low number of alleles per locus are typical of geographically restricted, endemic plants (Hamrick *et al.*, 1979). Small heterozygote deficiencies are often observed in populations of outbreeding species, and population substructuring (the Wahlund effect) usually is suggested to be the cause (Futuyma, 1986).

We provide a few diversity statistics in Table 6.3 to give an idea of the kinds of values one might calculate for different organisms. There is substantial differentiation among populations of kangaroo rats and among populations of autogamous plant species. Autogamous plants are highly inbred, amplifying the effects of isolation and genetic drift.

Species with restricted geographic ranges tend to have less total genetic diversity than widespread species and have most diversity within populations (Hamrick, 1983; Hamrick *et al.*, 1979). Species that are habitually inbred, such as autogamous plants, also tend to have reduced variability. Species restricted to a number of small, isolated remnants may have most genetic variation between populations (Sampson *et al.*, 1988). However, there are many exceptions and conservation geneticists are often surprised to find significant variation in small, isolated populations in which they fully expect relatively little variation (e.g. Schwaegerle and Schaal, 1979; Moran and Hopper, 1983, 1987; Coates, 1988).

Contingency χ^2 tests are available to test the significance of heterogeneity of gene frequencies among populations (see Workman and Niswander, 1970). The equations for genetic diversity used above may be biased if the sample size is relatively small. There is a more general formula for the indices of heterozygosity that accounts for sample size and for several polymorphic loci. Allelic evenness may be measured with an unbiased index given by the equation (Nei, 1973, 1975, 1978),

$$H_e = \frac{1}{m} \sum_{j=1}^{m} \left[2n_j \left(1 - \sum_{i=1}^{k} x_{i,j}^2 \right) \middle/ (2n_j - 1) \right] \qquad (6.6)$$

where $x_{i,j}$ is the frequency of the ith allele at locus j, summed over k alleles, and n_j is the number of individuals examined for locus j. The index is averaged over all m loci and it differs from H_T because monomorphic loci are included in the calculations.

Browsing through the literature on genetic diversity, it is easy to become confused by the terminology. The measure of evenness in a species (Equation (6.6)) goes by several other names including the effective number of alleles, expected panmictic heterozygosity, polymorphic index, and gene diversity (see Brown and Weir, 1983).

6.2.3   The language of genetic diversity

Genetic diversity is partitioned within species depending on the spatial distribution and reproductive behaviour of individuals. Thus, changes in gene frequencies due to genetic drift or the differential survival of genotypes may have ramifications at many organizational levels.

Description of genetic diversity needs to take into account the partitioning of variation among different components of a species. Population geneticists have developed a set of terms for the distribution of allelic variation that accounts for these difficulties and the terms can be arranged in a fashion analogous to Rabinowitz et al.'s (1986) classification for the terminology of the abundance and distribution of species. It is closely related to the notions of population abundance and distribution, and it emphasizes the fact that genetic diversity has two components, allelic richness and allelic evenness (Brown and Weir, 1983).

The geographic range in Figure 6.9 refers to the extent of the distribution of the entire species, without reference to any component of genetic diversity. A third category, regional, is sometimes included intermediate between widespread and restricted. A proportion of the alleles of any species will be common and a proportion will be rare within populations. This characteristic is guaged from the relative frequency of an allele within the species as a whole. Some alleles will be widely distributed among populations and others will be present in relatively few

Geographic range							
Widespread				Restricted			
Richness : allele frequency within populations							
common		rare		common		rare	
Evenness : allele frequency between populations							
Wide	Narrow	Wide	Narrow	Wide	Narrow	Wide	Narrow
1	2	3	4	5	6	7	8

Figure 6.9   A classification for loci by geographic range, allelic richness and frequency (after Marshall and Brown, 1975; Moran and Hopper, 1987). While the exact definition of these categories depends on the abundance and distribution of populations, in general, alleles are common if they occur in at least one population with a frequency >10%, otherwise they are rare. Alleles are widely distributed between populations if they occur in two or more populations, otherwise they are narrowly distributed. The numbers 1 to 8 are arbitrary labels for the classification.

populations. Thus, an allele may be fixed in a single, large population and therefore be relatively common. However, if it is absent from a number of other populations, its distribution may be quite narrow.

For a species where there are only two populations, the terminology is somewhat redundant, but *S. diclinis* makes an illustrative example (Table 6.1). The species is geographically restricted (Prentice, 1984). The fast and slow alleles of the EST-1 locus in *S. diclinis* may be relatively rare compared to the medium allele but all alleles occur in at least one population with a frequency greater than 10%, so all are considered common. The three alleles are widespread because they occur in both populations.

The distribution of genetic variation may be far more complex than in the case of *S. diclinis*. Different alleles may show different patterns of variability within and between populations. Some loci will drift randomly while others may respond to the pressures of natural selection. There may even be a hierarchical distribution of variation within populations, between populations within subspecies, and between subspecies, as Sampson *et al.* (1988) found for *Eucalyptus crucis*.

6.2.4   Genetic variation and population dynamics

Above, we have concentrated on descriptive measures of the genetic variation in natural populations. The measures are useful in addressing management questions such as which populations of a species should receive highest priority for conservation. For example, if a species exhibits a high degree of interpopulation differentiation, it may be necessary to conserve most or all the remaining populations if we are to conserve the full range of genetic diversity.

There is another important use for genetic information in conservation biology. It can be an invaluable indicator of the dynamics of fragmented populations, and a detailed example is provided by Templeton *et al.* (1990). They assayed isozymes from populations of salamanders (*Cryptobranchus alleganiensis*), collared lizards (*Crotaphytus collaris*) and lichen grasshoppers (*Trimerotropis saxatalis*) in the Ozark Mountains of the United States, and employed restriction site mapping of mitochondrial DNA to sample genetic variability when isozyme variation was low. They then overlaid the genetic surveys on the geographic distribution of the species.

In the species studied by Templeton *et al.* (1990), fusion of gametes is not mediated by a pollinator so the exchange of genetic material between isolated populations implies that individuals can move between populations. Significant differences in allele frequencies and the absence of shared haplotypes among populations is evidence of demographic isolation. Large amounts of within population variation is evidence that population sizes and reproductive behaviour are such that genetic drift is weak. Environmental variation will be the most important factor determining fluctuations in population size. If genetic variation within populations is small, it implies that demographic and genetic factors will be important in determining the persistence of the population, and the risk of local extinction is likely to be high.

This approach to the analysis of population dynamics in fragmented populations has a number of important advantages. Traditional dispersal studies rely on such things as capture, marking and recapture, or radio tracking. It is very difficult in studies of this kind to calculate the frequency of migration when it is a very rare event. A single migrant per generation may keep populations from diverging, and such an event could easily escape any traditional sampling design. Even if migrants move between populations, there is no evidence that they will contribute to the gene pool in the population in which they arrive. Genetic sampling circumvents all these problems, only at the expense of assuming that populations are at genetic equilibrium. We discuss this aspect of population genetics in the section below.

6.3 THE EFFECT OF INBREEDING ON POPULATION DYNAMICS

Before we go any further towards applying the ideas of population genetics to demographic risk assessment, we need to explore the notion of effective population size (N_e), and the measurement of inbreeding and migration. The equations to calculate the rate at which heterozygosity is lost from a population assume a randomly mating population in which there is no migration, mutation or selection. The most convenient way of dealing with populations that diverge from the idealized breeding structure is to express population size in terms of the effective population size. Inbreeding is important because there is a direct causal relationship between the degree of inbreeding, and the average values of survivorship and fecundity in many populations, a relationship we described above. To illustrate the use of the various equations presented below, we shall again draw on the examples of the white rhinoceros, for which we built a demographic model in Chapter 2, and on experiments with populations of *Drosophila* species.

6.3.1 Effective population size

The effective size of a population is the size of an ideal population that would undergo the same amount of genetic drift as the population under consideration (Wright, 1931; Kimura and Crow, 1963; Koenig, 1988). The best way to visualize an ideal population is as a chemostat, a soup in which a collection of individuals, identical except for sex, mix and reproduce at random. In particular, an ideal population has N individuals in discrete generations, all of whom are equally likely to contribute genes to the next generation. The next generation is a random sample of N male and female gametes and there is no correlation between fertility of parents and fertility of offspring. Thus, the genetic simulation model described in Section 6.2 above closely approximates an ideal population.

The rate at which variability is lost is not a function of the census number of individuals in a population. Rather, the mating system and reproductive behaviour of a species contribute to the rate of genetic loss. These factors must be taken into account when estimating chances of genetic deterioration. When the breeding structure is known, the effective size of a natural population can be derived from N, the census population size. However, to understand the breeding structure of almost any population sufficiently well to calculate N_e you will need very detailed information about the reproductive behaviour of the species as well as such things as the number of breeding individuals, the numbers in different age groups, the variation in lifetime dispersal of individuals, and the reproductive success of individuals of each sex.

The situation is complicated further by the fact that, despite the apparent simplicity of Wright's (1931) idea, there are three different kinds of effective population size, and numerous ways for estimating them that depend on the biology and mating system of the species under consideration, and the assumptions that can be made concerning reproductive behaviour. The three kinds of N_e are called the inbreeding, variance, and eigenvalue effective sizes.

Inbreeding effective size accounts for the probability of homozygosity by common ancestry, the amount of increase in homozygosity. It is the size of an ideal population with the same rate of decrease in homozygosity as the population under consideration. Variance effective size accounts for the amount of gene frequency drift per generation. It is the size of an ideal population with the same rate of increase in variance due to drift as the population under consideration. The eigenvalue effective population size accounts for the rate of loss of alleles at segregating loci. It is the size of an ideal population that has the same rate of loss or fixation of alleles as the population under consideration.

To read more about inbreeding and variance effective size, see Kimura and Crow (1963), Crow and Kimura (1970) and Crow and Dennison (1988). Ewans (1982) and Ewans *et al.* (1987) treat eigenvalue effective size in detail. The kind of effective population size you estimate depends on what aspect of the genetics of a population is most important. For example, when the question of interest is the increase in homozygosity due to random drift, inbreeding effective size is important. Variance effective size is appropriate when your interest is in the increase in variance among subpopulations or the rate at which genetic variation is lost from a population.

Effective population size is almost always less than the number of adults of reproductive age. The most important reasons usually are unequal numbers of males and females, variation in population numbers through time, or greater than binomial or Poisson variability in the number of offspring per parent (Crow and Dennison, 1988). Wright (1939), Kimura and Crow (1963), Lande and Barrowclough (1987) and Crow and Dennison (1988) have derived numerous expressions for inbreeding and variance effective population sizes, to account for different mating systems and reproductive behaviour. We reproduce a few of the equations that are likely to have the broadest applications.

Some of the equations described below assume diploidy, random mating, discrete generations, no selection, constant population sizes through time, and no correlation between the fertility of parents and the fertility of their offspring, unless they otherwise explicitly deal with these factors. Most of these assumptions will be violated by most natural populations, so estimating effective population size is a very difficult problem. Crow and Dennison (1988) derived equations to deal with numerous different mating systems if population numbers are not constant or if the variance in progeny number is not that expected from random mating. If we employ the simplifying assumptions described above, the variance and inbreeding effective population sizes reduce to the same equations.

One of the most important reasons effective population size is smaller than census size in many natural populations is that the number of males and females in the population differ. If this is so, then

$$\frac{1}{N_e} = \frac{1}{4N_m} + \frac{1}{4N_f} \tag{6.7}$$

where N_m and N_f are the census number of males and females, respectively. In the white rhinoceros population, there were in fact 18 adult males and 27 adult females (Conway and Goodman, 1989). Even though the ratio was not significantly different from 1:1, for the sake of illustration we substitute these values into Equation (6.7) giving an effective population size of 43. The slight imbalance in the sex ratio results in slightly elevated rates of genetic drift, resulting in the rate expected in a population of 43 individuals in which the sex ratio is 1:1.

Another important reason effective population size is smaller than census size in many natural populations is that there is variation in the number of progeny per individual. By chance, we would expect some variation but because of behavioural mechanisms, some members of a population may be far more successful than others. If a few individuals are inordinately successful, the chances are increased that alleles become homozygous by descent in subsequent generations. Effective population size accounting for differential reproductive success is given by,

$$N_e = \frac{4N - 2}{2 + \sigma_k^2}, \tag{6.8}$$

where σ_k^2 is the variance in the number of progeny produced per generation per individual, and N is the census number of individuals in the population. If a population has both unequal numbers of males and females, and variation in progeny numbers is different from random, then

$$\frac{1}{N_e} = \frac{1}{4N_{e,m}} + \frac{1}{4N_{e,f}} \tag{6.9}$$

where $N_{e,m}$ and $N_{e,f}$ are the effective numbers of males and females in the population respectively, and

$$N_{e,s} = \frac{N_s k_s - 1}{k_s + (\sigma_{ks}^2/k_s) - 1}. \tag{6.10}$$

The parameter k is the average number of progeny produced by an individual in its lifetime and the subscript 's' represents the sex, male or female (Lande and Barrowclough, 1987).

Conway and Goodman (1989) estimated a total population size of 45 adults and subadults in 1986. If we assume discrete generations and random mating in the rhinoceros population, the effective population size is 43. To provide an example of the effect of differential reproduction, we assume that 17 males each have an average of 1.67 offspring per generation and one male has 17 offspring. Each of the 27 females has an average of 1.67 offspring and any differences in success are due to random chance. We assume a constant census number of 45 adults in each generation.

Because mating is random among the females, N_{ef} equals the census number of 27 females. The variance in the number of offspring per male per generation is 13. Using Equation (6.10) we calculate that $N_{e,m}$ equals 6.6. Substituting these values into Equation (6.9) we estimate the effective population size to be about 21. This result says that if a single male is ten times as successful as other males in the population in terms of progeny per generation, the effective population size will be half the census population size. Genetic drift will occur in the rhinoceros population of 45 at a rate equal to that expected in a population of 21 randomly mating individuals.

Effective population size of the eastern barred bandicoot
The eastern barred bandicoot (*Perameles gunnii*) is widespread and abundant in Tasmania, and there are relatively large populations in the far eastern part of southern Australia. A population of the species near Hamilton in south-eastern Australia was isolated from other populations between 1937 and 1960. Sherwin and Brown (1990) estimated the population size to be:

census size: 633 ± 24

Sherwin and Brown (1990) then used the demographic and reproductive data at their disposal to estimate the effective size of the population. Because generations of the species overlap, they needed to account for the overlap in calculating N_e. Age-specific fecundity and mortality schedules are usually needed for these calculations (Lande and Barrowclough, 1987) but eastern barred bandicoots are difficult to age. Therefore Sherwin and Brown (1990) used an alternative approximation using the number of new recruits per generation (Crow and Kimura, 1972):

no. of new weanlings per generation: male   153
                                          female   340

Marsupials remain attached to their mothers for a period after birth until they are sufficiently developed to fend for themselves, so it was relatively easy to count the number of offspring to different females in the isolated population. The most surprising thing about the study was that two females shared more than 75% of the successful reproductive output in the population, whereas 11 females had no offspring at all (Figure 6.10).

From these data on female reproduction, they estimated the average number of progeny produced by an individual in its lifetime ($k_s = 1$) and the variance in the number of progeny produced per generation per individual ($\sigma^2_k = 11.6$). These values, substituted into Equation (6.10), gave:

corrected for lifetime reproductive success: males   24
females   54

Lastly, Sherwin and Brown (1990) accounted for the unequal sex ratio in the population using Equation (6.9), giving:

effective population size: 67

Thus the effective size is nearly one-tenth the estimated census population size. The rate of loss of genetic diversity from the population will be ten times higher than the rate of loss from the population had it behaved like Wright's ideal population.

Many natural populations do not have discrete populations but, as in the rhinoceros population, generations overlap one another. Overlapping generations present a special set of problems for the calculation of effective population size and there are many different formulas in the literature that make different assumptions about the structure and dynamics of populations.

Lande and Barrowclough (1987) provide equations to calculate the generation times of males and females, and the effective population size, using age-specific schedules of fecundities and survivorships. Reed *et al.* (1986, 1988) describe an analogous method: both approaches assume a stable age distribution, no immigration, stable population size, and no covariance between numbers of female and male offspring. An approximate formula that assumes the stable age distribution is provided by Hill (1979),

$$N_e = \frac{4N_cL}{\sigma_k^2 + 2} \tag{6.11}$$

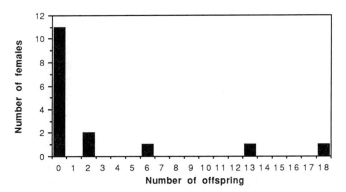

Figure 6.10 Lifetime production of pouch young by female *Perameles gunnii* at Hamilton, Victoria, Australia. (After Sherwin and Murray, 1990.)

where σ_k^2 is the variance in the number of progeny produced, and N_cL is the product of cohort size and generation length giving the total number of individuals per generation (see Falconer, 1989, pp. 72–6; cf. Crow and Kimura, 1972).

All population sizes vary and the effective size of a population depends on these variations in population size through time. If we have a census of the population in different generations, the effective population size is

$$N_e = \frac{1}{2}\left[1 - \left\{\prod_{i=1}^{t}\left(1 - \frac{1}{2}N_e(i)\right)\right\}^{1/t}\right]. \tag{6.12}$$

This equation can be approximated by the harmonic mean of successive population sizes, $(1/N_e = 1/t[1/N_1 + 1/N_2 + \ldots + 1/N_t])$, but only for short periods when gene frequencies are nearly constant (Lande and Barrowclough, 1987). Generations with small effective sizes will have a disproportionate effect on the long-term value of N_e, although when $N_e = 2$, at most 25% of the existing heterozygosity is lost in a single generation.

Individuals are not distributed continuously throughout their ranges and rarely is geographic proximity unimportant in determining mating patterns. Restricted gene flow, such as nearest neighbour pollination, reduces the effective size of a population and leads to rapidly increasing homozygosity and microgeographic variation (Turner et al., 1982). The degree to which individuals are isolated by distance is measured by the square root of the mean squared (r.m.s.) distance travelled by individuals from their birthplace to the point of first reproduction (see Koenig, 1988). For a population inhabiting a given area, for the purposes of the effects of random genetic drift, it may be considered panmictic if $ps^2 > 1$, where p is population density and s is the mean dispersal distance between an individual's birth site and its breeding site. If a population is not panmictic, we need to account for the neighbourhood size; Wright (1951), Maruyama (1977), Lande and Barrowclough (1987) and Koenig (1988) discuss different ways of doing so.

Perhaps the simplest way to estimate the effective size of a population is to calculate the rate of loss of heterozygosity from a population. Given two (or more) measurements of heterozygosity at different times from a single population, the effective population size can be calculated from (Crow and Kimura, 1970),

$$H_t = H_0\left(1 - \frac{1}{2N_e}\right)^t. \tag{6.13}$$

The advantage of this equation is that it subsumes all those factors of behaviour, demography and reproduction that contribute to the effective size of a population. The disadvantage is that for any relatively long-lived species, the rate of loss of variation may be difficult to measure.

The effective size of captive Drosophila *populations*
Briscoe *et al.* (1992) measured the rate of loss of genetic variation
from captive populations of *Drosophila melanogaster*. The eight
populations had been in captivity for periods ranging from 6 months
to 23 years. Variation was measured using a number of criteria
including allozymes, lethal frequencies, phenotypic variances and
additive genetic variances. Effective population size was estimated
from the rate of loss of variation in the populations using Equation
(6.13) above. It turned out to be between 3 and 5% of the census
population size. The effective sizes of two large populations founded
from wild stock were 1.6 and 0.4% of the census population.

In growing populations, variance N_e is likely to exceed inbreeding N_e
while in declining populations the reverse is likely to be true. When
population size changes rapidly, standard equations for effective population size may be very inaccurate (see Simberloff, 1988). None of the
equations for effective population size accounts for metapopulation
structure. The within and among population components of genetic variation and the effect of population size and migration on drift may have
important consequences for the estimation of effective size.

A great deal of information is required to estimate effective population
size in a natural population. Minimum requirements include demography,
dispersal, social systems and reproductive success. Obviously for many
natural populations, these data are not available. Franklin (1980), Soulé
(1980) and Lehmkuhl (1984) have summarized practical guidelines for
estimating N_e in the absence of critical information. For example, they
take the results of work on variance in number of progeny by Crow and
Morton (1955) and suggest that to estimate the effective population size,
the census number should be divided by a factor of about 1.4.

The example for the eastern-barred bandicoot above shows how such
generalizations may be quite inaccurate. In that case, we would need to
divide the census number by a value closer to 10. Nei and Graur (1984)
reviewed studies of allozyme variation in 77 species including plants,
mammals, fish, reptiles and snails. Effective population sizes were
very variable in relation to census size but were often less than 1%.
Avise *et al.* (1988) suggest from studies of molecular distances among
mitochondrial DNA lineages, that the long-term effective population
sizes of American eels, hardhead catfish and red-wing blackbirds are less
than 1% of current population sizes. All of these empirical studies suggest
that the rate of loss of genetic variability from captive or wild remnant
populations may be substantial, even when the census population size is
large.

6.3.2 Loss of heterozygosity and inbreeding coefficients

Inbreeding is a concept very closely related to the notion of genetic drift. The relative degree of inbreeding can be measured in natural populations and it is important in demographic studies because it has a direct effect on population mean values for survival and reproduction.

In bisexual species, every individual has 2^t ancestors, where t is the number of generations. It takes only a few generations before the number of ancestors exceeds any real population. Any two individuals must be related to one another through one or more common ancestors. The smaller the population size, the less remote are these ancestors, or the greater the number that a pair have in common. The consequence of common ancestors is that individuals may both carry replicates of one of the genes present in the ancestor. It is the production of homozygotes whose genes are identical by descent (autozygous) that gives rise to the increase in homozygotes as a consequence of inbreeding. The coefficient of inbreeding (F) is a measure of the relatedness of individuals, the probability that two alleles at any locus are identical by descent (Falconer, 1989, pp. 62–3).

Heterozygosity is reduced at a rate proportional to the inverse of the effective population size per generation (Wright, 1931; Crow and Kimura, 1970, p. 140). Inbreeding coefficients, F (Wright, 1931), are an index of the rate at which heterozygosity is lost from a population. The standard equation (Falconer, 1989, p. 71) for the rate of change per generation of the inbreeding coefficient for a population is

$$\Delta F = \frac{1}{2N_e + 1} \, . \tag{6.14}$$

This is a good approximation even for population sizes of the order of ten effective individuals. Conway and Goodman (1989) estimated the effective size of the rhinoceros population on Ndumu Reserve to be 33 individuals. Thus, ΔF per generation will be about 0.015. About 1.5% of the genetic variation present in the population will be lost per generation.

Inbreeding coefficients can only be interpreted with reference to a given population or generation and they measure the fraction by which heterozygosity is reduced per generation, relative to the reference population. We define a time (generation) in the past where we assume that all genes in the population are independent: it has a coefficient of zero. If the effective size of a population remains constant, the inbreeding coefficient of a population after t generations, relative to the initial population, is given by

$$F_t = 1 - (1 - \Delta F)^t. \tag{6.15}$$

Applying an arbitrary yardstick to the Ndumu rhinoceros population, we could aim to retain 90% of the variation presently in the population over the next 200 years. If the generation time of this rhinoceros population is less than 20 years, allowing only 10 generations in the next 200 years, using equation (6.15) we estimate that more than 10% of the remaining heterozygosity in the population will be lost. The effective population size is too small to avoid erosion of important amounts of the genetic variability of the population.

The effects of inbreeding in Drosophila *populations*
Borlase *et al.* (1992) used *Drosophila* as model organisms to test several common assumptions in conservation genetics. One of these is the assumption that higher levels of inbreeding will result in reduced levels of fitness in a population, and that the loss of fitness may be minimized by manipulating reproduction to increase the effective population size. They equalized the contribution of families to generations in small *Drosophila* populations, with the intention of increasing the effective population size in that line, and thereby reducing the rate of loss of heterozygosity. They allowed another line to persist in the same numbers through random mating. Ten replicate lines of each treatment were maintained using four pairs of parents per generation. After 11 generations, the mean inbreeding coefficient in the line with equalized family sizes was 0.341 and in the randomly mated line it was 0.498. The lines maintained with equalized family sizes had a relative fitness value of 0.260 after 11 generations compared to an outbred control line, whereas the line allowed to mate at random had a relative fitness of 0.123. The experiment suggests that equalization of family sizes may be a useful tool in maintaining reproductive fitness in captive populations.

Because inbreeding coefficients are an index of change in heterozygosity, they may be decomposed into different components representing among and within population genetic variability. Most commonly, isozyme data are used to estimate inbreeding coefficients, although they are best estimated from pedigrees (Crow and Kimura, 1970, pp. 67–73). There are three indices that measure inbreeding, accounting for the distribution of variance within and between populations. They are analogous to the measures of genetic diversity within and among populations outlined above. Wright's (1931) fixation index, F_{IS}, measures deviations of heterozygote frequencies from Hardy–Weinberg expectations. It is the probability that two homologous genes are derived from the same gene in a common ancestor in the same population. It is given by

$$F_{IS} = 1 - \frac{H_0}{H_e} \tag{6.16}$$

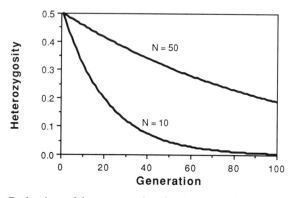

Figure 6.11 Reduction of heterozygosity due to genetic drift as a function of time (in generations). The curves were calculated using Equation (6.18). (After Chesser *et al.*, 1980.)

where H_0 is the average proportion of loci in an individual that are heterozygous, and H_e is calculated from Equation (6.6). The index ranges from $+1$ to -1, where positive values indicate a deficiency of heterozygotes. It measures the heterozygosity of individuals in a population relative to that of an individual derived from random mating in the same population (Crow and Kimura, 1970, p. 170).

Two other indices for inbreeding are commonly defined: F_{ST} is the probability that two homologous genes, chosen at random in a subpopulation, are both descended from a gene in the subpopulation. F_{ST} is approximately equal to the among group heterozygosity of a species, G_{ST} (Nei, 1973). It may be measured using Equation (6.4). Crow and Kimura (1970, p. 108) provide an alternative method for estimating it independently of G_{ST}. In a single population, this statistic estimates the overall inbreeding coefficient.

F_{IT} is the overall probability of identity by descent of alleles in an individual. The overall probability of identity is given by

$$F_{IT} = 1 - (1 - F_{IS})(1 - F_{ST}). \qquad (6.17)$$

Brown and Weir (1983) give details of methods to estimate each of the three inbreeding coefficients directly from gene frequency data that take into account the number of populations and varying sample sizes.

The inbreeding coefficient for a population is a function of the effective population size and the coefficients in preceding generations. Assuming, for example, that the number of males and females in a population is equal and there is no selfing, then (Crow and Kimura, 1970, pp. 320–1),

$$F_t = \frac{1}{2N_e} + \left(1 - \frac{1}{N_e}\right)F_{t-1} + \frac{1}{2N_e}F_{t-2}. \qquad (6.18)$$

By inspecting Equation (6.18), we can see that if a population goes through a period of inbreeding, a bottleneck or founder event, and afterwards expands quickly, the inbreeding is not undone but remains where it was before the population expansion (see Falconer, 1989, p. 64). Effective population size is critical in determining the rate at which variability is lost (Figure 6.11).

6.3.4 Migration

Genetic differentiation among populations is affected by migration. The frequency of an allele in a population is a balance between its frequency in the previous generation and its frequency among immigrants,

$$q_1 = m(q_m - q_0) + q_0 \tag{6.19}$$

where a large population consists of a proportion, m, of new immigrants each generation, q_0 is the frequency among natives and q_m is the frequency among migrants (Falconer, 1989, p. 24).

If we assume the number of populations is large and the mutation rate is small, then the relationship between differentiation and migration is given by (Wright, 1951; see Maynard-Smith, 1989)

$$G_{ST} = \frac{1}{1 + 4N_e m} \tag{6.20}$$

where m is the migration rate per population per generation, and N_e is the effective population size. If the level of between population differentiation is known, Equation (6.20) may be used to estimate the number of immigrants to a population. The equilibrium level of genetic differentiation between populations depends on the product of the population size and the migration rate, not the migration rate alone (Figure 6.12). If this product is greater than about 5, there will be little genetic differentiation, at least when there are many populations.

Figure 6.12 shows that the migration rate must be small if differentiation is to be important. By assuming populations are outbreeding, randomly mating and that there are no complicating chromosomal mechanisms in place, several authors conclude that one migrant per generation is sufficient to keep allele differences small between subpopulations (Wright, 1951; Lewontin, 1974, p. 213; Franklin, 1980; Frankel and Soulé, 1981; Ralls et al., 1983). Varvio et al. (1986) found in most cases when the number of migrants is one per generation, populations behave as if close to panmictic, even though the theoretical justification for the rule is questionable. For details of three methods for estimating gene flow including the F_{ST} (G_{ST}) method described above, the rare alleles method and a maximum likelihood method, including an evaluation of their advantages and disadvantages, see Slatkin and Barton (1989).

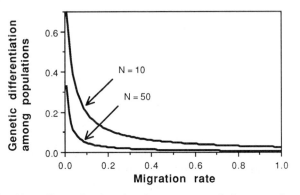

Figure 6.12 The effect of migration between populations on the equilibrium level of genetic differentiation among populations (G_{ST}). Migration rate is in units of the number of migrants per population per generation.

6.3.5 Equilibrium assumptions

Most of the equations outlined above assume that populations are in equilibrium. This assumption should be approached with care because extant populations are not necessarily in genetic equilibrium. This is especially true in areas where land has been cleared extensively in recent times. Furthermore, genetic systems may conserve heterozygosity, others may be neutral, while others may favour homozygosity. Intense selection against homozygotes in plants with abundant seedlings is a common phenomenon. In these species, selection counteracts the effects of inbreeding in small populations (e.g. Schaal and Levin, 1976; Brotschol *et al.*, 1986). In *Eucalyptus pendens*, a restricted small tree species with disjunct populations in land recently cleared for agriculture in Western Australia, there is no correlation between population size and any single locus measure of genetic diversity (Moran and Hopper, 1987).

The non-equilibrium view, appropriate for short-term dynamics, is to ask how long it may take to lose variation by random drift, what the maximum tolerable loss is (Lande and Barrowclough, 1987), and what effect these losses may have on population survivorships and fecundities. If our focus is on long-term dynamics, then equilibrium questions are appropriate. We can ask how much genetic variation is maintained in a population. Both views are likely to be important for wildlife management (Lande and Barrowclough, 1987).

Models for the loss of rare alleles by genetic drift assume there is no density-dependent selection on these alleles. There is ample evidence that at least some are alleles are maintained in balanced polymorphisms because they confer a selective advantage at low frequencies (see the review by Clarke, 1972). Fitness of genotypes is likely to change as the environment, allele frequencies and population densities change.

Frequency-dependent selection is likely to obscure relationships between heterozygosity and fitness.

Genetic variation in Eucalyptus lateritica

E. lateritica is a small, multi-stemmed (mallee form) eucalypt. It was discovered and named in 1986 and is confined to a range of 30 km in an area of lateritic hills near the coast of south-western Australia. It is known from ten discrete populations, each of 1–100 plants, usually located on the upper scree slopes and edges of flat-topped mesas. Isozymes were assayed from a number of the populations and measures of genetic diversity within and between populations (H_S, H_T, G_{ST}) were calculated. Total genetic diversity in the species was high ($H_T = 0.318$), and most of the genetic variation was found within populations, with relatively little differentiation between populations ($G_{ST} = 12.6\%$) (Moran and Hopper, 1987). These characteristics are more like those expected of large, outbred populations. This result is even more surprising considering eucalypts are pollinated by animals and it is expected that gene flow may be limited between isolated populations, increasing the degree of diversity between populations. However, the results are only surprising if we assume the genetics of the population are close to equilibrium conditions in a system without selection or genetic mechanisms that actively retain variation. Mallee form eucalypts are very long lived and have generation times of hundreds of years. Changes in the size and demographic structure of populations may occur on much shorter time scales. The genetic composition of the species may reflect much larger populations long since lost, or it may reflect selective or chromosomal mechanisms that are not apparent in allozyme assays.

Metapopulation dynamics may be important in determining gene frequencies in a species as a whole. The effects of local extinction and recolonization depend on how groups of colonists are formed, and on the numbers of individuals colonizing vacant habitat relative to the number entering extant populations (Slatkin, 1987; Wade and McCauley, 1988). These factors may also have a bearing on the fitness of individuals in a population, and therefore on the chances of persistence of a population. For example, Couvet *et al.* (1985) found that the ratio of females to hermaphrodites in populations of thyme (*Thymus vulgaris*) could not be explained if single populations were examined. Characters that are disadvantageous in a single population are frequent because of the advantages they confer in founding new populations, resulting in apparent disequilibrium in individual populations.

Lastly, the suggestion that one migrant per generation will be sufficient to keep a collection of subpopulations close to panmixia is derived from the equilibrium behaviour of G_{ST}. Populations are assumed to be at equilibrium. While G_{ST} is more likely to be close to its equilibrium value than H_S or H_T, all three may be very slow in equilibrating, depending on the number and size of the subpopulations (Varvio et al., 1986). There are few data on the temporal stability of genetic structure, or how it is related to population growth or demographic change over time.

6.4   A STOCHASTIC MODEL FOR *BANKSIA CUNEATA*

In this section, we develop a model for *Banksia cuneata* to illustrate the use of population genetics in a model to assess the risks faced by a natural population. *Banksia cuneata* is a large shrub native to the south-western part of Western Australia. It is rare and geographically restricted: there are about 340 individuals confined to seven isolated populations with a total range of 60 km (Taylor and Hopper, 1988). *B. cuneata* is characterized by two distinct genetic groups and Hopper and Coates (1990) suggest that a minimum of one large population from each of the two population groups would probably ensure adequate protection of the genetic resources of the species. Burgman and Lamont (1992) developed a stochastic model for a single population with the intention of predicting the effects of various management practices on the risks of population decline. The model illustrates the use of inbreeding coefficients for demographic risk assessment.

The populations of *B. cuneata* occur on road verges, small nature reserves and other State Government land about 180 km east of Perth in Western Australia. Individuals grow to 5 m tall in localized stands of sclerophyllous scrub-heath or low open woodland on undulating yellow sands (Lamont et al., 1991). The climate is dry Mediterranean and the mean annual rainfall of 347 mm falls almost entirely in winter.

Banksia seeds are produced in woody follicles forming cones. *B. cuneata* retains these cones and stores the majority of its seeds in the canopy. Like a number of other banksias, it is killed by fire; regeneration then depends on release of the canopy stored seeds (see Lamont and Barker, 1988). Seeds germinate the grow vigorously in ash beds created by fire. The valves are bound together by resin which melts when exposed to fire, allowing the valves to reflex and the seeds to disperse. Lamont et al. (1991) investigated the demography of one of the populations and noted several life history characteristics that would favour a vigorous population. The seeds of *B. cuneata* are rarely damaged by seed predators, and seed viability is high and remains more or less constant in time. The consumption of seedlings by herbivores such as rabbits, locusts and moth larvae is negligible. If fires occur in autumn, seeds released from burned adults have a good chance of avoiding granivory or desiccation and heat death over summer.

The summer drought kills numerous seedlings that establish after a fire. If the drought is extreme, most seedlings die. Mortality of seedlings during the first year is a direct result of competition with other seedlings and surviving adults for water (Lamont *et al.*, 1991). In a few of the populations, naturalized weed species dominate the ground storey and compete vigorously with *B. cuneata* seedlings for water following fire. Weed infestation reduces the chances of survival of banksia seedlings. Post-fire recruitment is critical. The interaction between soil preferences, drought stress and interspecific competition for water is probably responsible for the limited geographic distribution of many banksia species in south-western Australia (Cowling *et al.*, 1990, and references therein).

The crucial phase in the survival of individuals occurs during their first year. After fire, less than 5% of seeds germinate and less than 1% of these seedlings may survive the first summer drought. Through observations on a natural population and transplant and water supplementation experiments on seedlings, Lamont *et al.* (1991) concluded that population numbers are not limited by the size and dynamics of the seed canopy bank but by the weather pattern following fire-induced seed release. The limiting factor is water availability. Seedlings compete with each other, with adults and with other species for water during the root establishment phase.

There is reason to presume that long-term persistence of *B. cuneata* in the wild is in doubt. The distribution of individuals in seven localized stands in a region with a short wet season makes the species vulnerable to repeated fire, habitat loss and extended drought. Adults and seedlings are killed by fire and young seedlings are very susceptible to the annual summer drought. Since the late 1970s, the populations have been monitored and they have been either approximately constant in numbers, or declining (Western Australian Department of Conservation and Land Management, unpublished data).

6.4.1 Genetic parameters

In the sections above, we described the causes and consequences of inbreeding depression. We noted that the relationship between the loss of heterozygosity – Wright's inbreeding coefficient (Wright, 1931) – and reduction in fitness of the offspring has been shown in empirical studies to be nearly linear for many species of mammals, insects and plants: a 10% increase in the inbreeding coefficient results in approximately a 10% deterioration in characters directly related to reproduction and survival. Burgman and Lamont (1992) used this generalization to estimate the reduction in recruitment in *B. cuneata* that occurs as a result of inbreeding.

The change in the inbreeding coefficient per generation is a function of effective population size. The simplest assumption is that the plants in a population mate at random and that there is random variation in the number of offspring produced by different individuals. Burgman and Lamont (1992) assumed the *B. cuneata* population was close to Wright's (1931) idea of an ideal population in which the change in the inbreeding coefficient per generation (ΔF) is given by Equation (6.14). These assumptions may result in overestimates of the effective population size if there is a significant proportion of self-pollination in the population. Other rare, bird-pollinated plants growing in the same part of Western Australia have low effective outcrossing rates (e.g. *Eucalyptus rhodantha*, Sampson *et al.*, 1989).

D. Coates (personal communication) sampled seedlings and adults from several of the *B. cuneata* populations at different times, and estimated levels of heterozygosity in the populations from isozyme studies. He found intense selection against homozygotes in the seedling populations. Selection against homozygotes in plant populations with abundant seed is a common phenomenon (e.g. Schaal and Levin, 1976; Brotschol *et al.*, 1986; Tigerstedt, 1988). As a result, we may expect more heterozygotes and a smaller rate of loss of heterozygosity than the neutral model predicts (i.e. the population will be more outbred than the corresponding population in the absence of selection). However, when dealing with traits closely associated with fitness, such as reproduction and survival, there is usually little additive genetic variance on which selection can act. The effect of selection on recruitment is very likely to be small, and we can ignore it in the model. The equation for the change in the inbreeding coefficient per generation is given by Equation (6.14). The effective population size is assumed to be the same as the number of adults in the population.

The demographic data were collected by Lamonet *et al.* (1991) from a population growing on a road verge. There were 79 living plants in the population ranging in age from 2 to 31 years in 1989. A total of 23 individuals were sampled from the same population by D. Coates (personal communication). He measured allelic heterozygosity in the parental population and in the progeny and calculated inbreeding co-efficients for the two generations. He found F for the parents was 0.190 and F for the progeny was 0.181. The change in the inbreeding coefficient from parents to progeny was not significantly different from zero. There were 47 adults alive in the population at the time he sampled, so we expect from Equation (6.21) that the change in the inbreeding coefficient should have been less than 0.001, a value too small to be detected given his sample size (23 plants) and the variability of the sample. D. Coates did detect significant increases in inbreeding coefficients in two other populations of the species.

6.4.2   An age-structured model including fire, rain and inbreeding

Germinated seeds have little chance of surviving the seasonal drought, even in a wet year. Plants aged 1–4 years differ fundamentally from older plants in not producing seeds. Plants aged 5–12 normally survive summer drought but differ from older plants in having relatively few seeds stored in their canopy. Plants older than 12 years have very little chance of succumbing to seasonal drought, and produce a superabundance of seeds after fire. Plants older than 20 years have increasing chances of mortality due to branch splitting.

Burgman and Lamont (1992) typified the individual plants in stands as belonging to one of 45 age classes reflecting differences in fecundity and tolerance to drought and fire. They assumed the population is censused in May, after drought and fire have had their effects, and just before germination in winter. They used an age-structured model for population growth (see Chapter 4) with an annual time step because germination is confined to a very short period in winter when rainfall is adequate. All important events are mediated by the strong seasonality of rainfall conditions.

Because of the superabundance of seeds on plants more than 12 years old, Burgman and Lamont (1992) assumed the number of seeds is never limiting in plants older than 12 years. That is, the number of recruits is independent of the number of viable seeds in the canopy of each individual. The dispersal radius of seeds from adult plants is small, so that competition for water will be important even when the adult population is very small. Therefore, the number of recruits is proportional to the number of adults. In years of average rainfall following a fire, there are about 1.5 new recruits for each adult older than 12 years, and 0.75 new recruits for each adult between 6 and 12 years old (Lamont *et al.*, 1991; B. Lamont, unpublished data). The higher value for larger plants reflects the fact that they are taller and produce more seed, allowing them to disperse further and hence more seedlings can establish. Burgman and Lamont (1992) repeated many of the analyses below with a continuous function of fecundity and obtained results that were qualitatively the same.

Competition for limited water is important among seedlings. When seedling densities are of the order of 100, survival can be as high as 40% after one year (Lamont *et al.*, 1991). However, under natural conditions after fire, densities are always of the order of thousands per square metre and the number of survivors per adult is a function of rainfall. That is, Burgman and Lamont assumed a constant dispersal area around each adult. Within that area, the number of recruits is determined by rainfall. The species usually grows in dense thickets and appears to be adapted to high population densities because adult mortality due to branch splitting is higher among individuals that grow in isolation. The density of individuals in the seven stands is far below the carrying capacity of the environment. Given a net positive growth rate, the populations could increase to many times their current size (B. Lamont, unpublished data). For these reasons, Burgman and Lamont (1992) omitted a carrying capacity for adult plants.

Recruitment in the absence of fire is rare and sporadic. New recruits were observed to survive at a rate of 0.01 recruits per adult per year, for adults five years and older (Lamont et al., 1991). This mean is used as an estimate of the recruitment rate when there is no fire. The number of new recruits is further modified by inbreeding depression. Burgman and Lamont (1992) assumed an inverse, linear relationship between the inbreeding coefficient, F, and the number of recruits, whereby $m_{t+1} = m_t(1 - \Delta F_t)$ for years in which there is a fire, and $m_{t+1} = m_t(1 - \Delta F_t/G)$ in years when fire does not occur. The generation time, G, can be estimated from the schedule of survivorships and fecundities (Burgman and Lamont, 1992), and standard Lotka equations (see, for example, Krebs 1985, p. 189). The generation time for the population is approximately 23 years. Taking all of the above factors into account, there were two equations for the number of new recruits each year:

$$N_1(t + 1) = \sum_{x=5}^{12} N_x(t)m_j(t) + \sum_{x=13}^{45} N_x(t)m_a(t) \qquad (6.21)$$

$$N_1(t + 1) = \sum_{x=5}^{45} N_x(t)m_k(t) \qquad (6.22)$$

where $N_{1,t+1}$ is the number of new recruits at time $t + 1$, $N_{x,t}$ is the number of adults in age class x at time t, and m_j and m_a are the number of new recruits per juvenile and adult plant respectively. The term m_k represents the average number of offspring per plant in the absence of fire, and its value was 0.01. The term ΔF_t is the change in the inbreeding coefficient at time t given by Equation (6.14). Equation (6.21) applies in years when there is fire, and Equation (6.22) applies in years when there is no fire. In years when there is no fire, the number of plants in the other age classes is given by the equation

$$N_{x,t+1} = N_{x-1,t}\, p_x \qquad (6.23)$$

where p_x is the probability of surviving from time t to time $t + 1$. In years when a fire occurs, the number of plants in classes older than the first class are set to zero.

6.4.3 Environmental and demographic variability

Equations (6.21)–(6.23) describe a deterministic model for growth of a population of B. cuneata. Burgman and Lamont (1992) used mean observations to represent the parameters for fecundity and survivorship. However, there are a number of factors in the life history of the species that are inherently stochastic. Lamont *et al.* (1991) noted that the survivorship of seedlings depends on the amount of rain that falls during the winter after seeds have germinated. Mean annual rainfall is low, even by Mediterranean climate standards. As mean rainfall decreases from location to location in Western Australia, it becomes more variable. The distribution of annual rainfall figures is skewed to the right (Hall *et al.*, 1981; Geng *et al.*, 1986). As a result, there is a greater chance, in any one year, of observing less than average rainfall than there is of observing greater than average rainfall.

For simplicity and because of a dearth of empirical observations, Burgman and Lamont (1992) assumed a linear relationship between rainfall and the number of recruits following a fire. In Equation (6.22) the parameters m_j and m_a were set equal to $0.5(-4.77 + 0.018P)$ and $(-4.77 + 0.018P)$ respectively, where P is annual rainfall. Because the distribution of historical records for annual rainfall is skewed to the right, they chose P from a lognormal distribution, with a mean of 347 and standard deviation of 83, values based on empirical observations of rainfall for the region. These equations give 1.5 recruits per adult in years of average rain, and 0.6 recruits per adult when rain is 15% below average. The number of recruits increases linearly with increasing rainfall. If rainfall is very low, no seedlings survive.

There is no correlation in the data of Lamont *et al.* (1991) between the number of recruits in years without fire and the amount of rain in those years. Thus, recruitment estimated by Equation (6.22) was calculated independently of rainfall.

One management alternative for these populations is to burn them at regular intervals. To model controlled burning, Burgman and Lamont (1992) spaced fires evenly in time and the entire population was burned. In all cases where controlled burning was modelled, the first burn was in year five of the simulation.

B. cuneata populations are small and demographic stochasticity may be important in affecting population persistence. If each birth and death in the population is independent of other births and deaths, the number that survive and the number recruited into the population are a random sample whose variance depends on the population size and the age-specific fecundities and survivorships. In the model, Burgman and Lamont (1992) estimated the number of survivors by sampling the binomial distribution, and the number of new recruits by sampling the Poisson distribution (see Brillinger, 1986; Akçakaya, 1990b). When there is a fire, survivorships and fecundities are neither random nor independent. In any case, the variability due to demographic uncertainty will be swamped by the variability in rainfall. In the model, demographic stochasticity is ignored in these years.

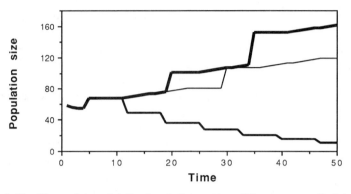

Figure 6.13 Three deterministic simulations using different controlled burning regimes. Fires at 15-year intervals are represented by the heavy line, fires at 25-year intervals are represented by the thin line, and fires at seven-year intervals are represented by the medium line. The other parameters are as described in the text. Equations (6.21)–(6.23) describe the population dynamics. (After Burgman and Lamont, 1992.)

We used the model developed by Burgman and Lamont (1992) and simulated population growth over a period of 50 years. The mean and 95% confidence limits of the total population size for each simulation were calculation from 1000 replications of each simulation. The smallest value obtained in each replication was also recorded and used to contruct a graph representing the chances that the population will fall as low as a given threshold size, at least once in next 50 years. The simulations were begun using the age distribution in the population reported by Lamont *et al.* (1991) for the initial distribution.

To establish the degree to which different parameters affected the performance of the model, Burgman and Lamont (1992) completed numerical sensitivity analyses. Each parameter in the deterministic model was varied and the corresponding change in population size was recorded. The model was run using fires at 15-year intervals as the standard against which to compare the changes induced by changes in parameters. The parameters were reduced by 10% and the effect of the changes was measured from the mean change in total population size over the period of the simulation (see Beck, 1983). They found the deterministic model was highly sensitive to survivorship values, and somewhat less sensitive to mean rainfall and fecundity values. The model was relatively insensitive to inbreeding.

6.4.4 Predictions

The deterministic model was run using three different intervals for control burns. Burning at 15-year intervals results in the highest predicted

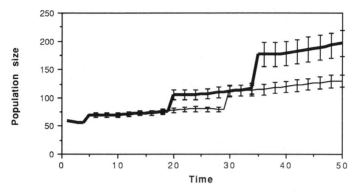

Figure 6.14 The stochastic simulation of the model including demographic stochasticity, variation in annual rainfall and controlled burns at 15 year (heavy line) and 25 year (thin line) intervals. The error bars represent the 95% confidence limits for the mean population size and the simulation for fires at seven year intervals has been omitted. The results are based on 1000 replications of the simulation. (After Burgman and Lamont, 1992.)

population size because 15-year-old adults are fully productive and it effectively increases the reproductive rate above burns at 25-year intervals (Figure 6.13). Burning every seven years reduces the expected population size because plants less than 12 years old carry relatively few seeds.

If we switch off the inbreeding parameter in the model by arbitrarily setting ΔF to zero, the effect on the deterministic model is small. The simulation based on fires at 15-year intervals reaches a population size of 172 after 50 years, compared to 161 when inbreeding is present. If we were to rely on the deterministic model, the results suggest that the best way to maximize the future size of the population is to burn at an intermediate frequency, once every 15 years.

The mean values for the stochastic simulations are qualitatively the same as those for the deterministic simulations. The first thing to note, however, is that confidence limits for the simulation based on fires at 15-year intervals are relatively broad compared to those for the simulation based on fires at 25-year intervals (Figure 6.14). The uncertainty in the predictions is due to the variability in rainfall and its effect on recruitment after fire. Superficially, it appears as though there is little need to be concerned for the persistence of the population because, in both cases, the mean population sizes and the 95% confidence limits are well above zero. However, recall from Chapter 2 that these limits may be misleading.

If inbreeding is ignored in the stochastic model for either scenario of fire regimes represented in Figure 6.14, there is no significant difference between the population size at the end of 50 years and the mean population size at the end of 50 years in the presence of inbreeding. The reason

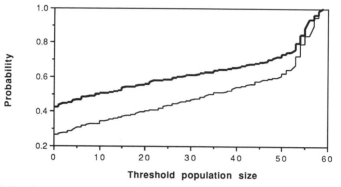

Figure 6.15 Quasiextinction risk curves for the population of *B. cuneata*. They give the chances that the population will fall below a specified threshold at least once in the next 50 years given fires at 15 years intervals (heavy line), and fires at 25 year intervals (thin line). The results are based on 1000 replications of the simulation.

is that 50 years, a little over two generations, is too short a time for inbreeding to have much of an effect on the population. Given an effective population size of, say, 50 individuals, only about 2% of the variation in the population will be lost in two generations. The reduction in mean population size and the increased chances of extinction of the population of *B. cuneata* associated with inbreeding are not detectable because they are masked by the environmental stochasticity in the system.

The quasiextinction risk curves for these simulations are interesting because they are not entirely intuitive (Figure 6.15). Even though the mean population size for the scenario in which the stand is burned every 15 years is higher, this practice results in elevated risks of population decline and extinction. The problem for *B. cuneata* essentially is that, after a fire, seedlings are faced with the seasonal summer drought. If insufficient rain falls, recruitment will not replace the adult population in place before the fire. Because of the lognormal distribution of rainfall in semi-arid Western Australia, years of below average rain occur more often than not. Even though the stochastic model predicts that the population will probably increase, there remains about a 40% chance that the population will become extinct within the next 50 years if it is burned every 15 years (Figure 6.15).

Inbreeding had no important effect on the risks of population decline. It did not matter if inbreeding was included in the model or not because the risk curves were unchanged for each of the simulations in Figure 6.15. We know from Equation (6.10) that the effects of inbreeding will be manifested most rapidly in small populations. To display a noticeable effect of inbreeding, it was necessary to modify the conditions of the

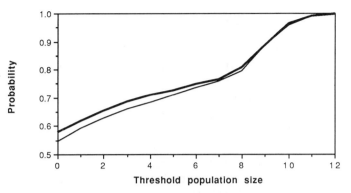

Figure 6.16   Quasiextinction risks, within 100 years, for the population of *B. cuneata* consisting initially of 10 plants. They give the chances that the population will fall below a specified threshold at least once in the next 100 years given fires at 25 year intervals in the presence of inbreeding depression (heavy line) and in the absence of inbreeding depression (thin line). The results are based on 1000 replications of the simulation.

simulation such that the *B. cuneata* population consisted initially of ten plants, and we ran the simulations over 100 years.

Given a smaller population and a longer time horizon, we expect to detect some effect of the reduction in fecundity implemented in Equations (6.21) and (6.22). After 100 years, the mean population size in the absence of inbreeding was 59.9 (95% confidence limits equal 11.2) and in the presence of inbreeding it was 49.0 (95% confidence limits equal 10.1), based on 1000 replications of the model. There was an effect of inbreeding on the risk of population decline, although it was slight (Figure 6.16). These results suggest that any factor that reduces population size will compound the risk of population decline by heightening the effects of inbreeding on the population.

6.4.5   Management implications

The focus of the management plan for this species is not to maximize the expected population size at the end of 50 years, but to minimize the risk of extinction over the next 50 years. The management procedures that optimize these two different aims are quite different. Burgman and Lamont (1992) investigated a number of other management options for the species and concluded that the most effective plan is to provide water for seedlings in the summer drought following fire, at least in years when rainfall is low. Lamont *et al.* (1991) have shown experimentally that seedling survival may be greatly improved by watering over summer. If this option cannot be implemented because of monetary or logistic constraints, the next best thing is to burn the sites at infrequent intervals when the seed store of the trees is high but before they begin to senesce.

These results are a qualitative guide at best. The parameters have been estimated from scant data and Burgman and Lamont (1992) do not suggest that the model results in trustworthy predictions for future population sizes. However, insofar as the fundamental dynamics of the model are a reasonable representation of the population dynamics of *B. cuneata*, the qualitative actions suggested by the results are our best estimate of the most efficient way to manage the populations.

The total population of *B. cuneata* is composed of seven separate stands. Their approximate sizes are known and they are far enough apart that fires may occur independently in each stand. Several of the existing stands are small, fewer than 20 individuals. In small stands, we can expect that inbreeding will have a detrimental effect on mean population size and the risks of population decline over the next 100 or more years. There is no migration of individuals between stands, but there is a chance that they exchange gametes if pollinators move between the populations. If there is exchange of genetic material between populations, it may have some effect on the dynamics of small populations by reducing the rate at which the inbreeding coefficient increases with time.

6.5 THE GENETICS OF METAPOPULATIONS

The total amount of genetic variation in a species is composed of genetic variation between individuals within populations, and genetic variation between populations. That is, genetic variation is partitioned among and within the elements of a metapopulation. Quantitative geneticists have worked out ways of estimating the equilibrium amount of variation expected in populations of different spatial configurations and sizes, with different rates of dispersal between them. The various models and their assumptions were detailed above.

A more immediate problem in conservation biology is the short-term, non-equilibrium dynamics of small and fragmented populations, and the impacts of these dynamics on the risks of extinction. The complexity of natural populations and the unique nature of many of their behavioural and ecological mechanisms makes it necessary to build explicit models of these populations. The development of simulation models for the genetics of metapopulations has lagged behind the development of models for population dynamics, partly because of the complexity of the problems involved, and partly because the fate of genetic diversity in a metapopulation depends on the dynamics of the population. It was necessary for good population models to develop before work could begin on good genetic models for the same systems.

Gilpin (1991) has taken the problem in hand and developed a computer program to simulate genetic drift in a metapopulation. He uses an occupancy model for the dynamics of the metapopulation, in which populations inhabiting local patches have extinction probabilities proportional to the carrying capacity of the patch. Each diploid individual has ten loci (genes), each with two alleles. States are assigned randomly to the alleles (represented by a 0 or a 1), such that the initial allele frequency at each locus is 0.5.

The model works by assigning constant probabilities of dispersal between patches, and constant extinction probabilities within patches. Because it is an occupancy model, a recolonised patch goes from having no individuals to having a population at the carrying capacity in a single time step. Thus, the genetic composition of a migrant that recolonizes an empty patch becomes that of the entire population in a single time step, and as a result the model generates a pronounced founder effect. The model also ignores distances between patches. Instead, Gilpin makes colonization probabilities proportional to the carrying capacities of the source local population. He allows one random recombination event per genome per time step.

The simplifications that Gilpin (1991) makes are necessary because of the number of parameters involved in a model with full spatial structure. In such a model there will be all the parameters of the appropriate occupancy model for a given number of patches, as well as parameters and variables representing the genetic make-up of individuals within each population.

6.5.1 A model for two populations of white rhinoceros

The most restrictive assumption in Gilpin's system is that patches are either empty or fully stocked. The obvious solution is that the genetic model should be linked to a metapopulation model that includes the dynamics of populations within patches (see Chapter 5). In the model for white rhinoceros developed in Chapter 2, we included only females and we modelled the dynamics of the population by simulating exponential growth to a ceiling, the number to which the population is limited by the reserve managers. To provide an example of the genetics of metapopulations, we develop a model for two such populations, including both males and females. We use it to calculate the rates of loss of heterozygosity and estimate the impacts of migration between the populations.

We specify a carrying capacity of 50 individuals for each patch. Populations grow exponentially whenever they are below the carrying capacity and are truncated when they reach it. At each time step, each adult has a 0.08 chance of dying and each female has a 0.28 chance of producing an offspring. We assume that the census is taken just after reproduction so that juvenile mortality is negligible, and we ignore age structure. Migration takes place after reproduction. We assume there are no detrimental effects of outbreeding, although such a phenomenon could easily be included in the model. The details of the model are provided in the algorithm below.

Algorithm 6.1

1. Specify the probability of dispersal E, between the populations, the correlation between environments, and the standard deviations of birth rates b_i and death rates d_i, for each population, i.
2. For each replication, do steps 3 to 11.
3. Initialize population sizes, the sexes of individuals and their genetic arrays.
4. For each time period, t, do steps 5 to 10.
5. Select the birth and death rates for time t for each population, i.
6. For each female, choose a random number U from the uniform distribution. If $U < b_i$ and $N_i < 51$, that female produces an offspring. If a birth occurs, choose randomly one of the two gametes from that female, allow one recombination event with the other gamete, and assign the gamete to one half of the genome array for the offspring. Choose randomly one male from the population and repeat the procedure to assign a gamete to the genome array of the offspring.
7. For each member of each population, choose a random number U from the uniform distribution. If $U < d_i$, that individual dies.
8. For each population i, choose a random number U from the uniform distribution. If $U < E$ and $N_j < 50$, choose a migrant from i and move it to j.
9. Record the population size at t, the current time period.
10. Calculate H_S, H_T and G_{ST} at t from the genome arrays.
11. Record minimum and maximum values for H_S, H_T and G_{ST} for this replication.
12. Calculate average population sizes (over all replications) at each time step; calculate average H_S, H_T and G_{ST} values and the probability of their decline from the records of minimum and maximum H_S, H_T and G_{ST} values.

An array is associated with each individual representing 20 loci on a single chromosome, each with two alleles. Initially, each of the alleles in the populations has a frequency of 0.5 and the states are assigned randomly among individuals. Sex is assigned randomly to each newborn. The newborn individuals take one gamete from their male parent and one from their female parent and one recombination event is allowed between the gametes of the male and female contributing to the newborn individual. Migration will only occur from population i to population j if there is a vacancy in population j, that is, if N_j is less than the carrying capacity. Once births, deaths and dispersal events have been recorded

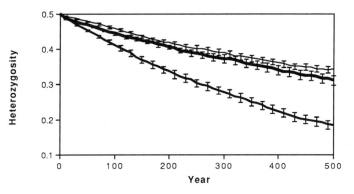

Figure 6.17 Heterozygosity within and among populations of white rhinoceros. With no migration between subpopulations, mean heterozygosity within populations (medium line) and total metapopulation heterozygosity (thin line) for two population of white rhinoceros. When one migrant per year is exchanged between population, total heterozygosity and mean within population heterozygosity are more or less identical (heavy line). The vertical lines represent the 95% confidence limits for mean heterozygosity.

and the new genetic arrays specified, the quantities H_S, H_T and G_{ST} are calculated as outlined above.

In the first set of simulations, there was no migration, and in the second, one migrant per year moved between the populations. The dynamic model includes removal of individuals when the populations exceed 50 individuals, and populations fluctuate because of demographic and environmental variation. Correlation of environmental fluctuations between the two populations was 0.1. The upper, thin line in Figure 6.17 represents the total genetic heterozygosity in the entire metapopulation when there is no migration between the two populations. Mean heterozygosity within populations is given by the lowermost line on Figure 6.17. The result indicates that, in the absence of migration, the populations become fixed for alleles, resulting in a decline in mean heterozygosity within populations. The rate of decline is much higher than in a metapopulation in which there is migration. However, because of the vagaries of genetic drift, when the two populations do not exchange migrants, they become fixed for different alleles and the total genetic variability is higher than if migration occurs.

We may quantify the impact of migration on heterozygosity in this example by employing risk curves. The minimum values for each quantity H_S and H_T were recorded in each replication of the simulations and were used to construct curves representing the chances that heterozygosity will fall below specified levels (Figure 6.18).

The risk curves appear in an order that may have been predicted from the mean values in Figure 6.17. When populations are isolated from

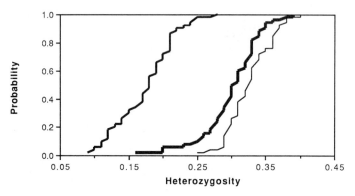

Figure 6.18 Risk curves for mean heterozygosity within populations (medium line) and total metapopulation heterozygosity (thin line) for two population of white rhinoceros between which there is no dispersal. For both total and mean heterozygosity (heavy line) for the populations in which 1 migrant per year moves between them, the risk curves are identical.

one another, within population heterozygosity is almost certain to fall below 0.25 within the next 500 years, whereas total metapopulation heterozygosity is very unlikely to reach that level. If one migrant per year is exchanged between the populations, there is about a 15% chance that heterozygosity will fall below 0.25.

6.5.2   Genetic variation and vital rates

There is a lack of information for all but a handful of species concerning the crucial connection between genetics and demography. In the *Banksia cuneata* example, we assumed a linear relationship between heterozygosity and reproduction. In the rhinoceros example with two populations, we assumed that loss of genetic variability had no effect on either the mean or the variability of rates of fecundity or survival.

These assumptions represent two very simple genetic systems. There are an enormous number of such systems in higher organisms, governed by the chromosomal mechanisms operating within the genome, the size and distribution of populations, and the reproductive behaviour of individuals within populations. The assumptions we have made above are unlikely to be true for many species, and yet the relationships between genetic variability and the vital rates of a population will be critical in determining both the rate of loss of heterozygosity and the chances of persistance of populations and species.

It is unlikely that the relationship between genetic variability and vital rates will have a dramatic impact on most populations over the short term (10–100 years), depending on the generation time and the effective population size of the species. However, for a genetic model to be generally useful in answering questions for conservation biology, it must predict not only the expected genetic changes but also how these changes will affect the population's viability through their effects on survival, fecundity, dispersal and so on.

Perhaps more importantly, reproductive management strategies in conservation should seek to maintain the genetic structure of populations. In particular, the break-up of locally coadapted gene pools by the introduction of immigrants may result in the breakdown on these complexes, the expression of recessive lethal genes and rapid declines in the vital rates of a population. The use or avoidance of interpopulational dispersal and hybridization in the management of conserved species in fragmented habitats should be based on a thorough understanding of the genetic systems involved (James, 1982).

Thus, in a sense, the simulations presented above are misleading because, particularly in the case of the white rhinoceros, the model is based on little direct evidence of the genetic system it seeks to model. It would be unwise to make generalizations concerning the effects of various management practices such as the translocation of individuals between isolated patches from such models. As with all other kinds of models, there is no substitute in population genetic models for thorough knowledge of the biology of the species concerned.

6.6 SUMMARY

Loss of genetic variability may effect the ability of a species to adapt to environmental change and it may reduce population average survivorships and fecundities. Genetic drift is a stochastic process which results in the loss of alleles from a population. The average rate of loss of alleles is a function of population size.

Usually, genetic diversity is sampled from patterns of phenotypic variation observed among isozymes. Isozyme variation for a single locus in the rare shrub *Silene diclinis* suggests little genetic differentiation between two subpopulations. Overall levels of genetic diversity in the species are typical of geographically restricted, endemic plants.

The distribution of genetic diversity within and among populations (termed allelic richness and evenness, respectively) is important in determining the rate at which genetic diversity is lost from populations. The chances that an allele will be lost is higher in small populations, as is the reduction in heritable quantitative variation. By sampling genetic variation in fragmented populations, it may be possible to evaluate demographic characteristics and migration patterns of natural populations. This approach is especially useful in detecting rare migration events.

Effective population size is the size of an ideal population that will undergo the same amount of genetic drift as the population under consideration. It is necessary to estimate effective population size if the rate of genetic deterioration is to be calculated. It depends on the breeding system of the species, including such things as the sex ratio, the number of progeny per adult, variation in the total population size over time, and overlapping generations. The way in which effective population size is calculated may depend on what aspect of genetic deterioration is important. In most cases in conservation biology, either inbreeding effective size or variance effective size will be appropriate.

Inbreeding depression results from genetic drift in small effective populations. There is a close relationship between inbreeding in populations that are usually outbred, and the vitality and fecundity of individuals. Most methods for estimating the loss of genetic diversity from a population assume that gene frequencies have reached some sort of equilibrium. The white rhinoceros population, for which we developed a model in Chapter 2, is likely to lose significant amounts of genetic variation over the next 200 years if no unrelated individuals are introduced from other areas.

A stochastic, age-structured model for *Banksia cuneata* includes reduction in fecundity as a function of the size of the population. The effects of inbreeding on expected population sizes and the chances of quasiextinction in the short term (50–100 years) are unimportant in large populations, those with more than about 50 effective individuals. In small populations, with fewer than about ten effective individuals, there is an important effect of inbreeding on population size and the risks of quasiextinction over periods as short as four generations (about 100 years). The short-term expectations of survival of *B. cuneata* populations depend more critically on patterns of rainfall following periodic fire than they do on genetic factors, which may be more important to long-term adaptation to environmental change.

Gilpin's (1991) work indicates the direction in which models for the genetics of metapopulations may develop in the future. An example is developed for two populations of white rhinoceros along the lines he proposes. It demonstrates the role that migration may have in partitioning genetic variation within and among populations.

To predict the loss of heterozygosity from metapopulations, a thorough knowledge of the dynamics of the populations and of their genetic systems is needed. The parameters, assumptions and details included in any such model should be tailored to the species under consideration. Selection against young individuals with relatively high levels of homozygosity is a common phenomenon in plants, and could be modelled with a function that reduces survivorship as a function of the proportion of homozygous loci. Any chromosomal or behavioural mechanisms that impinge on the genetic effective population size should be accounted for in estimating the loss of heterozygosity from the metapopulation. It is important to have some idea of the effect of changes in genetic variability on the mean and variance of survivorships and fecundities.

There are many aspects of conservation genetics we did not address in this chapter. Probably the most important one is the use of pedigrees in very closely managed populations (Lande and Barrowclough, 1987; Falconer, 1989). There are other things relevant to natural populations that we have ignored. For example, the genetic diversity of a natural population may be profoundly affected by its age structure (Crow, 1979), or by the release of epistatic genetic variance and the subsequent increase in additive genetic variance that may result from a founder event (Goodnight, 1988). Just as for models that focus on the demography of populations, when trying to forecast the genetic status of a population we should always be prepared for surprises. Either of these factors could result in unexpected changes in the genetic composition of a population.

In both of the examples we developed above, the genetic variability of the populations are not at risk in the short term. Both theoretical and empirical evidence suggest that demography presents a more immediate and substantial risk that does the loss of genetic variability (Lande, 1988b). The focus of conservation genetics should be to understand the interactions between genetic structure and variability, and demographic parameters such as reproduction and survival.

7 Extensions of risk assessment

7.1 THE SCARCITY OF DATA

All but the simplest models in this book defy parametrization because of a lack of data. This underlines the complexity of superficially simple questions such as 'What increase in risk results from this management alternative?' or 'Is it better to spend money on the rehabilitation of a corridor, or on the extermination of a feral predator?' It also underlines the scarcity of data essential to answer some of the most fundamental questions facing conservation biologists.

In an ideal situation, there are many kinds of information that could be used to assess the risks faced by a population or metapopulation. They include the following.

1. Demographic data: the size of the population, and age-, stage- and sex-specific fecundities and survivorships.
2. Behavioural data: the breeding, feeding, migratory and territorial behaviour of individuals at different population densities and as a function of their demographic attributes.
3. Habitat data: the components of the environment that make up the niche of a species, the distribution and quality of habitat and changes in availability and quality of habitat in space and time.
4. Functional and physiological data: the causal relationships between environmental variables and demographic parameters.
5. Spatial data: the distribution of individuals, their patterns of movement relative to the distribution of their habitat, and the demography, behaviour and ecology of different groups.
6. Genetic data: levels of inbreeding, the structure of coadapted gene complexes, the distribution of genetic variation within and among populations, and the quantitative effects of inbreeding and outbreeding on demographic parameters.
7. Stochastic data: the variability of age-, stage- and sex- specific fecundities and survivorships, the patterns of variability of environmental variables that determine births and deaths in a population, and correlations between changes in these demographic parameters and environmental variables with time and in spatially separate patches of habitat.

This list is not exhaustive. In particular, it ignores interactions with other species such as predation, mutualism and competition. Even though we would like information on all these aspects of a species' biology, often all we have is an estimate of the current size of the population, limited distribution data and a few guesses at some of the other parameters and variables based on cursory observations, anecdotal evidence, and extrapolations from so-called similar species. These problems are compounded by the fact that validating stochastic models used for risk assessment requires very large data sets that provide information on both the averages and the statistical properties of the state variables.

Because of the imminent threat of extinction faced by many species, conservation biologists are forced to make judgements and recommend management alternatives in the absence of relevant and complete data, and with the assurance that no such data will soon be collected. As a result, all management plans for conservation are constrained by shortages of personnel and resources, and by conflicting priorities in the use of the natural environment. However, to some extent, any difficulties concerning the absence of data are superfluous. Building a model within realistic constraints is nothing more than rationalization of a problem and the quantification of ideas. Even though the models generate quantitative predictions, common sense tells us to trust only the qualitative result, even within the framework of risk assessment. A large number of characteristics could be included in any model of a natural population and we choose to include only those that we expect to be important. Any predictive model may very quickly come unstuck because a critical parameter has been omitted.

No matter how many data are collected, there will always be a trade-off between reality (detail) and tractability. There is a fine balance between models that closely reflect reality but are impossibly complex, and models that are unreal but possible. The danger is that simple models will do no more than reflect one's preconceptions. On the other hand, models that are more complex than the data allow could do the same thing. There is little value in replacing an unrealistic structure with an unmeasurable parameter.

The fact that only rarely is even a small proportion of the necessary data available in any single case does not reflect on the strengths or otherwise of risk assessment for conservation biology. An analysis based on the most scant data will serve at least to clarify different ideas and highlight the components of the data that are pertinent to the most critical questions.

7.2 SPECIES VIABILITY ANALYSIS

We define Species Viability Analysis as the analysis of all those factors and their interactions that act on *species* and contribute to the risks of short- and long-term decline or extinction. While many problems in conservation biology are related to single populations, consideration needs to be given to the viability of species: usually species are the objects of our conservation efforts. As the examples in Chapter 5 demonstrate, optimal management strategies for minimizing risks in single populations may not be the same as those that minimize the risks in the metapopulation as a whole. Just as important, what is good for one species may not be good for others.

Species Viability Analysis necessarily includes not only the interactions of spatially seperate populations, but also the interactions of a species with other species in its environment. Conner (1988) suggests that populations of organisms should be conserved in sufficient abundance to be ecologically functional in their community, whereas the common approach, and the one we outline in this book, is to evaluate the risks faced by populations and metapopulations without considering the communities in which they live. All the methods we describe in this book ignore the functional role of populations and the interdependence of species. Such analyses add a further level of complexity to the kinds of analyses we have described, and will demand more data and better understanding than we have currently for most species. While a comprehensive Species Viability Analysis is beyond the limits of the data for perhaps any species, elements of it could be included in a model where the knowledge exists. For example, if an important component in the dynamics of a species is the abundance of a predator, there is nothing to preclude a function for it in the analysis.

Analyses of the viability of species that account for community interactions will almost certainly result in findings that are different from those that result from studies that focus on populations, or even metapopulations. For example, the population studies in the preceding chapters show that increased variability in the environment leads almost exclusively to heightened risks of quasiextinction. This generalization is at least partly the result of the myopic view we take of populations. Chesson and Warner (1981) found that variability in the environment in both space and time has the potential to enhance the chances that a species will persist in a community where competition between species and competitive exclusion are important processes. Species that are, on average, poorer competitors and that would be excluded in a uniform environment will have an advantage in some places at some times in a variable environment. Occasional good fortune may be enough to allow these species to persist indefinitely.

The process for building a model for risk assessment, be it for a population, a metapopulation, or a species, should be an iterative one. The first step is to design the structure and make guesses about functional relationships before data collection begins. The design should be such that the model will be able eventually to furnish answers to relevant questions. The detail in the model at this early stage should reflect how much is known or can be guessed about the species' behaviour and dynamics, and the constraints of time and the ease of data collection that are anticipated. Then, a feedback relationship should be employed between the structure and complexity of the model, the relative importance of the different parameters to the model's predictions, and the difficulty (cost) of collecting the data. The model can be changed and the sensitivity of the predictions to the various parameters re-evaluated as new knowledge is aquired. The methods developed in this book will work equally well, whether the model is for a single population or for an assemblage of interacting species.

7.3   IMPLEMENTATION

There are two generalizations that can be made from viability analyses that are important for the implementation of management plans. First, the size of a viable population is not static (Usher, 1987). As we emphasized in the discussion of the classification of species into categories of threat in Chapter 1, changing circumstances will alter the size of a population needed to provide an acceptable chance of persistence over a given time frame. Estimates of the threat faced by a species based on risk assessment models will change as new information is found and is incorporated into the model. The population sizes necessary to achieve acceptable levels of risk will vary for different species at different times and in different places.

Second, the factors that seem to be important in the predictions of average population sizes may be partially independent of the factors that determine the risks of population decline and extinction. A factor may increase average growth rate of a population and thereby increase expected population sizes, but it may also cause an increase in the variance in growth rate and thereby increase the risks of extinction. People who manage populations may need to manage for both average population size and extinction risks independently, trying to maximize the former and minimize the latter.

Apart from a relative handful of wildlife managers based in government institutions around the world, most ecologists have had little to do with the management of the populations they study and less to do with decision making and policy formulation within the bureaucracies that administer wildlife management. For this reason, much ecological research has little direct relevance to many of the immediate and pressing problems facing wildlife managers.

Lindenmayer et al. (1993) have reviewed wildlife conservation policy processes and characterize them in terms of a series of predictable steps originally defined by Brewer and deLeon (1983). These steps involve the identification of a problem, definition of the problem using expert analysis and technical systems, formulation and authorization of a response, implementation of a programme, appraisal of the implementation effort, and finally, discontinuation of the programme and revision of policy. Lindenmayer *et al.* (1993) emphasize that often the various steps in the process are the responsibility of different people, in different agencies and with different agenda.

Risk assessment methodology may be used in all of the processes. As Lindenmayer *et al.* (1993) point out, analyses of the viability of populations will allow a rigorous and repeatable definition of problems. Policy response and implementation may be guided by estimates of the impacts on risk of different management options. Different programmes may be compared and their efficacy measured in terms of their effects on expected population sizes and risks of population decline.

There are many statistics that can be applied to problems in conservation, including expected population sizes, extinction and quasiextinction risks, and the distribution of the time to extinction. These measures provide different pictures of a population and its future. When designing strategies to conserve or manage biological species, one must choose which to use as a basis for decisions. For example, a strategy that maximizes average population growth will not necessarily maximize the time to extinction. Conservation may be carried out with different objectives in mind. Were we to proceed using Conner's (1988) suggestion that species be preserved in such numbers that their ecological role is maintained, we would often specify a higher minimum threshold than if our intention was to do no more than preserve at least some members of a species in the wild. The index one uses should reflect the intention of the conservation effort.

Methods for risk assessment may be usefully applied in decision-making processes through the specification of 'acceptable' levels risk. This involves a judgement that is, by definition, anthropocentric and which depends inevitably, at least to some extent, on the cultural and economic values of the species under consideration. Burgman and Neet (1989) emphasize that it is essential to distinguish the tool, the analysis of the risks of extinction, from the object, the conservation of species. The analysis of extinction risks permits us to make decisions and modify our actions in a manner that minimizes the risks of extinction. The question of which species to study and why, remains.

Given that risk assessment provides a quantitative basis for policy formulation and conservation planning, ecologists should tailor their research efforts so that they address the immediate needs of planners and managers. Hobbs and Hopkins (1990), in reviewing policy needs for environmental planning, suggested that land-use management procedures should be more adaptive to cope better with environmental uncertainties. This recommendation echoes the sentiment of Holling (1973; Lindenmayer *et al.*, (1993), and Hilborn (1987) who suggested the development of systems that accommodate future events no matter how unexpected these events are. Uncertainty is not restricted to variability in the environment but may be part of our changing attitudes towards a problem. For example, changes in the organization of bureaucracies that manage natural areas, changes in levels of funding, or changes in public sentiment, may determine what are considered to be 'reasonable' management alternatives.

Environmental planning will be improved by the integration of research and management. In the words of Lindenmayer *et al.* (1993), viability analyses should be integrated with, not substituted for, other approaches to wildlife management. Ecological research that aims to establish critical parameters and functional relationships for viability analysis will be focused on answering questions with direct applications for managers. The iterative development of models for viability analysis provides a vehicle for adaptive management (in the sense of Holling, 1973). New information may be evaluated and incorporated in the structure of the model to provide revised estimates of the risks faced by populations. Uncertainty about future events is encapsulated in these estimates of risk. We hope the framework of risk assessment developed in this book will allow the models to be couched in terms useful for answering questions in conservation biology.

Appendix

A.1 RANDOM NUMBER GENERATORS

When we write a computer program to implement a model of a natural population, we need a source of random variation to represent the variation inherent in natural systems. For example, we may sample uniform random numbers to decide if individuals in a population have survived. If the chance of survival from one year to the next is 0.8, we sample the uniform distribution for each individual in the population. If the number is greater than 0.8, the individual dies, and if it is less than or equal to 0.8, the individual survives. Random numbers from the normal and lognormal distributions are used in a similar way to represent the effects of the environment on a population. A number of books are available that provide background to numerical simulation including Knuth (1981, pp. 114–40), Press *et al.* (1986) and Ripley (1987). Swartzman and Kaluzny (1987, pp. 176–208) give details of many methods useful for the stochastic simulation of population dynamics, including hints and examples on programming style and the modelling of environmental variables.

One of the most important tools in numerical simulation is a random number generator. Computer programs that represent population dynamic models may be written in any of a number of different programming languages. All random number generators in common usage, those available in all the most widespread programming languages, do not produce sequences of true random numbers. Rather, they produce very long sequences of numbers from deterministic equations. There is a period after which the numbers they generate will repeat in exactly the same order. They are termed pseudorandom number generators and the most important thing about them is that they must produce a deterministic sequence of numbers with the same relevant statistical properties as a sequence of true random numbers if they are to be useful in simulations. Sequences of pseudorandom numbers are sufficiently long and complex to behave just like true random numbers for most purposes. However, Ripley (1987) points out that pseudorandom number generators may have serious defects and it is important to test the generator at your disposal for the properties most important to the type of simulation you perform.

The vast majority of generators produce a set of numbers such that each value between 0 and 1 has an equal chance of being selected (the uniform distribution). When modelling population growth, a common process is to ask, 'is a random number larger or smaller than a given value?' such as in the example of determining if an individual dies or survives a period of time, and then to take alternative action depending on the outcome of the question. Thus, one important property of pseudorandom numbers is uniformity. A simple test for uniformity is to divide the interval between 0 and 1 into a number of equally spaced classes (say, 10), then choose a large number of random numbers ($>10^6$) and count the frequency of numbers in each of the classes. The uniformity of the pseudorandom number generator may be tested by comparing the observed frequencies with the expected (i.e. equal) numbers in each class.

If there is some correlation between the magnitudes of adjacent numbers in the sequence of pseudorandom numbers, it could have important consequences for the simulation of population growth. If the correlation is negative and you are using the numbers to represent environmental conditions, you will get alternate good and bad years more often than you should by pure chance. To test the independence of pseudorandom numbers in a sequence, runs tests may be employed to test if more or fewer series of increasing or decreasing numbers are observed than could be expected by chance alone. For these and other tests of independence and uniformity, see Ripley (1987) and Swartzman and Kaluzny (1987).

A.2 ALGORITHMS FOR RANDOM NUMBERS

It is beyond the scope of this book to provide complete details of all the methods for all the statistical distributions you may need to simulate the dynamics of populations. We have used the normal, lognormal, uniform, binomial and Poisson distributions in various examples in the text, and the computer code to sample them may be long and complex. Happily, whole textbooks are devoted to the topic of numerical simulation and Press *et al.* (1986) provide computer code to sample many statistical distributions. The code they provide is thoroughly tested and may be used in recipes for subroutines within a program without needing to understand the complexities of the underlying logic. The code is written in the computer languages FORTRAN or PASCAL, because these languages are relatively widespread and are well known. An alternative to using the computer code given in textbooks is to write to your own code, based on algorithms. Below, we provide some simple ways for generating random numbers for a few of the most important distributions.

Even when using algorithms from recommended texts, it is always a good idea to check that they behave properly. For example, you may write a subroutine using the code supplied by Press *et al.* (1986) to generate normally distributed random numbers. The numbers it generates are used to represent random variation in the environment. If you pass two values to the subroutine representing the mean and variance, it will return a random number from a theoretical distribution with the specified mean and variance. If you use the routine to generate many such numbers (again, $>10^6$), there are numerous tests that may be applied to determine if the distribution returned to you has the same mean and variance as the values you supply to the program. See Sokal and Rohlf (1981) for tests of normality, t-tests and so on.

A.2.1   The binomial distribution

The most common use for this distribution is to simulate the number of individuals that survive in a population from one year to the next, given a mean survivorship probability for the population (Akçakaya, (1991). The binomial distribution is relatively easy to sample for small numbers of objects. The best way is to use what is known as the 'direct' method, and we used this approach in the algorithms including demographic stochasticity in Chapter 2. Simply choose a number from the uniform distribution and ask whether it is larger or smaller than the critical probability, p. Take one of the two alternative actions (such as, the animal lives or the animal dies) depending on the outcome. The question may be asked and the uniform distribution sampled for each member of the population at each time step. This method is very simple to implement and works quite efficiently in terms of computer time for populations of less than about 50. If the population is very much larger than that, more complex algorithms to sample the binomial are available in the texts referred to above.

A.2.2   The Poisson distribution

The most common use for this statistical distribution is to determine how many offspring are produced in a population, given an average number of offspring per individual (Akçakaya, (1991). Ripley (1987, p. 55) gives a simple and efficient algorithm to sample a Poisson deviate, X, for a population with a mean value, μ (Algorithm A.1).

Algorithm A.1   *A method for sampling the Poisson distribution*

1. Set $P = 1$, $N = 0$, and $c = e^{-\mu}$
2. Repeat steps 3, 4 and 5 until P is less than c.
3. Sample a number from the uniform distribution, U.
4. Let $P = P \cdot U$.
5. Let $N = N + 1$.
6. Let the Poisson deviate, X equal $N - 1$.

Note that you don't need to supply the variance of the distribution because the variance is equal to the mean.

A.2.3 The normal distribution

There are many algorithms available to generate normally distributed random numbers. The most common use of normal deviates in ecological risk analysis is to represent random variation in the instantaneous growth rate r of a population, and variation in survivorships and fecundities that results from variation in the environment. The 'polar' method for normal deviates (Knuth, 1981, p. 117; Ripley, 1987, p. 62) is one of the most efficient and widely implemented methods (Algorithm A.2). It returns two independent, normally distributed variables, Y_1 and Y_2, sampled from a normal distribution with a mean value μ and a standard deviation σ.

Algorithm A.2   The polar method for sampling the normal distribution

1. Sample two random numbers from the uniform distribution, U_1 and U_2.
2. Let $V_1 = 2U_1 - 1$.
3. Let $V_2 = 2U_2 - 1$.
4. Compute $S = V_1^2 + V_2^2$.
5. If S is greater than or equal to 1, return to step 1.
6. Compute $X_1 = V_1\sqrt{\dfrac{-2\ln S}{S}}$.
7. Compute $X_2 = V_2\sqrt{\dfrac{-2\ln S}{S}}$.
8. Compute $Y_1 = \mu + \sigma X_1$.
9. Compute $Y_2 = \mu + \sigma X_2$.

One drawback that limits the efficiency of this algorithm is that it produces two random numbers, even though you may only need one. The superfluous value can be retained until it is required. More efficient, and more complex, algorithms are provided by Press *et al.* (1986).

A.2.4 The lognormal distribution

The lognormal distribution is often used to represent variation in parameters that are known to be skewed to the right, and to represent variation in the finite rate of increase of a population, R, that results from variation in environmental conditions. Usually, a random number from a normal distribution such as those produced by Algorithm A.2 is used in the procedure to generate a random number from the lognormal distribution. One such algorithm is provided in Algorithm A.3 and it returns a number, L, sampled from the lognormal distribution with a mean μ and a standard deviation σ.

Algorithm A.3   *A method for sampling the lognormal distribution*

1. Let $c = \sigma/\mu$.
2. Compute $m = \ln(\mu) - 0.5 \ln(c^2 + 1)$.
3. Compute $s = \sqrt{\ln(c^2 + 1)}$.
4. Sample a random number, Y, from the normal distribution with a mean m and standard deviation, s.
5. $L = e^Y$.

Occasionally, the random number you wish to generate has natural limits, such as the restriction that survivorships must be between 0 and 1. The lognormal is bounded on the left by 0, but there is always a chance the resulting number will be greater than 1. It wouldn't make any sense to have survivorship probabilities greater than 1: a population can't do any better than have everyone survive. The usual procedure is to truncate any values greater than 1 so that they are euqal to 1. However, this will have an effect on the mean and the variance of the values that are obtained from the sampling procedure. The mean will be biased downwards and the variance will be reduced as a result of truncating the values. The effect will be particularly strong if the probability of survival is close to 1. A large proportion of the random numbers will exceed the limit and be truncated.

One way to minimize this problem is to subtract from unity any survivorship, p, greater than 0.5, then sample a random number from the lognormal distribution using the value $(1 - p)$ for the mean. When the random number L is obtained, revert back to the original scale by again subtracting the value from unity (i.e. $1 - L$). This procedure has the effect of reflecting the lognormal distribution, and the number of values that exceed 1 will be reduced.

A.3   RANDOM EVENTS AND CORRELATED RANDOM NUMBERS

You may want to include specific details of the effect of the environment on a population in a model for demographic risk assessment. We did this in the model for *Banksia cuneata* in Chapter 6 when we included the effects of prescribed burning on the population at regular intervals. If these events are both mutually exclusive (either there is a flood or there is a fire) and random, it is possible to specify the probabilities of the different events and include these chances in a model. Suppose that in any single year it floods with a probability $p1$, there is fire with a probability $p2$, there is destruction due to human activity with a probability $p3$, and there is an equitable year free of disturbance with a probability $p4$. To decide which of these events occurs in any one year, we may generate a random number, U, from the uniform distribution and specify (see Knuth, 1981),

Flood	if	$0 \leq U < p_1$
Fire	if	$p_1 \leq U < p_1 + p_2$
Disturbance	if	$p_1 + p_2 \leq U < p_1 + p_2 + p_3$

This process is entirely analogous to using the direct method to sample the binomial distribution. The only difference here is that we have more than two possible events and the population may follow one of several paths, depending on the outcome.

A.3.1   Arrival times

Environmental events such as catastrophes may be included in a model as events that differ in some qualitative way from the variability in population growth rates, that is, the sum of the effects of many environmental parameters on the population. For example, the size of an isolated mammal population may vary because the winter is colder or longer from one year to the next, and we represent this kind of variation adequately with random variation in the population growth rate parameter. However, we may wish to include in the model the unlikely possibility that the population will be devastated by disease. The chance of an immigrant may be rare, and the chance that the immigrant carries a disease is still rarer. Two other statistical distributions will help us to model these kinds of events.

If an event occurs randomly with a frequency of once every μ time steps, then the time between two successive events has the exponential distribution with mean μ (Knuth, 1981, p. 128). To sample a number, X, from the exponential distribution with a mean, μ, compute $X = -\mu \ln U$. Using this approach, you will be able to specify the time at which the next event occurs, and avoid having to sample the uniform distribution at every time step.

Similarly, if some event occurs with probability p, the number of N independent trials needed until the first event occurs (or the number between occurrences) is given by the geometric distribution (Knuth, 1981, p. 131). To sample a random number from this distribution, let $N = \ln U / \ln(1 - p)$, and round the resulting value to the nearest integer.

A.3.2   Correlated random numbers

Occasionally, data on the response of a population to different environmental variables suggest that two random parameters are correlated; that is, the effect of these variables on the population is only partially independent. It's important, then, to be able to generate correlated random numbers. A method to generate correlated random numbers is given in Algorithm A.4 (Ripley, 1987):

Algorithm A.4   *A method for generating correlated random numbers*

1. Generate X_1 and X_2, independent normal deviates with $\mu = 0$ and $\sigma = 1$.
2. Let ρ be the required correlation coefficient between the two variables.
3. Compute $Y_1 = \mu_1 + \sigma_1 X_1$.
4. Compute $Y_2 = \mu_2 + \sigma_2(\rho X_1 + X_2\sqrt{1 - \rho^2})$.

The values Y_1 and Y_2 will be normally distributed random variables with means μ_1 and μ_2, standard deviations σ_1 and σ_2, and with a correlation coefficient ρ between them. If there are more than two correlated variables, as in a Leslie matrix, we may want to specify the standard deviation of each and its correlation with the other elements in the matrix.

The methods for calculating these values are more complex than in the case of two variables. Correlated random variates can be computed as weighted linear combinations of independent random variates. For example, let $z_1, z_2, \ldots, z_n$ be independent standard normal deviates. We desire to construct n new variables $y_1, y_2, \ldots, y_n$ which have the variance–covariance matrix

$$
P = \begin{bmatrix}
\sigma_{11} & \sigma_{12} & \cdots & \sigma_{1n} \\
\sigma_{21} & \sigma_{22} & \cdots & \sigma_{2n} \\
\cdot & \cdot & & \cdot \\
\cdot & \cdot & & \cdot \\
\cdot & \cdot & & \cdot \\
\sigma_{n1} & \sigma_{n2} & \cdots & \sigma_{nn}
\end{bmatrix}
$$

The correlates obtained as

$$
y_i = \sum_{j=1}^{i} c_{ij} z_j, \quad i = 1, \ldots, n
$$

where c_{ij} are elements of the lower triangular matrix C such that $CC' = P$. The elements of C may be determined by the following 'square root method' (see Wold, 1955; Moonan, 1957; Faddeeva, 1959; Scheuer and Stoller, 1962; Knuth, 1981, p. 551):

$$
c_{i,1} = s_{i,1}/\sqrt{S_{11}}, \quad 1 \leqslant i \leqslant n
$$

$$
c_{ii} = \sqrt{s_{ii} - \sum_{k=1}^{i-1} c_{ik}^2}, \quad 1 < i \leqslant n
$$

$$
c_{ij} = \left(s_{ij} - \sum_{k=1}^{j-1} c_{ik}c_{jk}\right)\Big/ c_{jj}, \quad 1 < j < i \leqslant n
$$

$$
c_{ij} = 0, \quad i < j \leqslant n
$$

A.4   MORE ABOUT SENSITIVITY ANALYSIS

Once a model has been constructed, we need to know how sensitive are its predictions to our estimates of each of the parameters. We can determine this by varying the parameters by a small amount around the estimated values. If, for example, a 10% increase in a parameter results in a uniform 100% change in the predicted population size, the state variable (population size) is sensitive to the parameter.

In Chapter 3, we provided one very simple method for numerical evaluation of the sensitivity of a model to its parameters and initial conditions. The method is appropriate for deterministic models and does not provide a complete picture of the sensitivity of the analogous stochastic model to its parameters. There are, furthermore, many other ways in which the sensitivity of a model may be evaluated. An alternative approach is to use the sum of squared deviations of a simulation from a standard trajectory (for an example, see O'Neill *et al.*, 1980). If values that diverge markedly from the standard are to weigh much more heavily than values that diverge only marginally, squaring the deviations will provide the desired result. It is analogous to using squared deviations to fit regression models in analysis of variance problems. The sensitivity of the variable L_i to the parameter P_j will be defined as

$$S_{i,j} = \frac{1}{n} \sum_{t=1}^{n} \frac{(\mathrm{d}L_{i,t}/L_{i,t})^2}{(\mathrm{d}P_j/P_j)^2}$$

The approach to measuring sensitivity above was developed for deterministic models. The problem with stochastic models is that it is difficult to know how much the observed deviations are due to the deterministic effect of the parameter, and how much they are due to random variation between the sets of replications. Swartzman and Kaluzny (1987, pp. 217–34) provide details of a number of methods for analysing the sensitivity of stochastic models, and list four characteristics common to good methods:

1. The method should be clearly defined.
2. The effects of interactions between parameters must be distinguishable from single parameter effects.
3. The method must include information on the variability associated with parameter estimates.
4. The method must allow interpretation of several output variables.

One way of addressing the problem of the sensitivity of stochastic models is to make use of a characteristic of most random number generators in computers. If they are provided with the same seed, they generate the same sequence of random numbers. Thus, we can run sets of replications for a stochastic model, and then repeat the whole simulation with the same pseudorandom numbers, after changing the parameters slightly.

All we need do is provide the same seed for the random number generator in our computer, and ensure the program samples the same number of random numbers, no matter what the outcome of the simulation (see Section A.1).

If the state variable of interest is the mean population size, the variable L becomes the mean population size. The deviations, dL, are the differences between the mean population size using the standard parameter values, and the size using altered parameter values. It will mean that we shall have to sample the variability of the parameters in some fashion and run the program many times. The sensitivity of the model to a parameter will be the mean of the values for S that result from the simulations.

In addition to the state variable, N, one may also analyse the sensitivity of quasiextinction risk to the values of parameters.

References

Akçakaya, H. R. (1990) Bald ibis *Geronticus eremita* population in Turkey: an evaluation of the captive breeding project for re-introduction. *Biological Conservation*, **51**, 225–37.

Akçakaya, H. R. (1991) A method for simulating demographic stochasticity. *Ecological Modelling*, **54**, 133–6.

Akçakaya, H. R. (1992) Population viability analysis and risk assessment, in *Proceedings of Wildlife 2001: Populations* (ed. D. R. McCullough), Elsevier, Amsterdam.

Akçakaya, H. R. and Ferson, S. (1990) *RAMAS/space User Manual: Spatially Structured Population Models for Conservation Biology*, Applied Biomathematics, Setauket, New York.

Akçakaya, H. R. and Ginzburg, L. R. (1991) Ecological risk analysis for single and multiple populations, in *Species Conservation: A Population-Biological Approach* (eds A. Seitz and V. Loeschcke), Birkhauser Verlag, Basel, pp. 78–87.

Allee, W. C. (1931) *Animal Aggregations: A Study in General Sociology*, University of Chicago Press.

Allee, W. C., Emerson, A. E., Park, O., Park, T. and Schmidt, K. P. (1949) *Principles of Animal Ecology*, Saunders, Philadelphia.

Allen, K. R. (1980) *Conservation and Management of Whales*, University of Washington Press, Seattle.

Allendorf, F. W. and Leary, R. F. (1986) Heterozygosity and fitness in natural populations of animals, in *Conservation Biology: The Science of Scarcity and Diversity* (ed. M. E. Soulé), Sinauer, Massachusetts, pp. 57–76.

Alvarez, L. W., Alvarez, W., Asaro, F. and Michel, H. V. (1980) Extraterrestrial cause for the Cretaceous-Tertiary extinction. *Science*, **208**, 1095–1108.

Andrewartha, H. G. and Birch, L. C. (1954) *The Distribution and Abundance of Animals*, University of Chicago Press, Chicago.

Avise, J. C., Ball, R. M. and Arnold, J. (1988) Current versus historical population sizes in vertebrate species with high gene flow: a comparison based on mitochondrial DNA lineages and inbreeding theory for neutral mutations. *Molecular Biological Evolution*, **5**, 331–44.

Ayala, F. J. (1968) Genotype, environment, and population numbers. *Science*, **162**, 1453–9.

Ballou, J. and Ralls, K. (1982) Inbreeding and juvenile mortality in small populations of ungulates: a detailed analysis. *Biological Conservation*, **24**, 239–72.

Baltensweiler, W. and Fischlin, A. (1988) The larch budmoth in the Alps, in *Dynamics of Forest Insect Populations: Patterns, Causes, Implications* (ed. A. A. Berryman), Plenum Press, New York, pp. 331–51.

Barkham, J. P. (1980) Population dynamics of the wild daffodil (*Narcissus pseudonarcissus* L.) I. Clonal growth, seed reproduction, mortality and the effects of density. *Journal of Ecology*, **68**, 607–33.

Bartlett, M. S. (1960) *Stochastic Population Models in Ecology and Epidemiology*, Methuen, London.

Bayliss, P. (1989) Population dynamics of magpie geese in relation to rainfall and density: implications for harvest models in a fluctuating environment. *Journal of Applied Ecology*, **26**, 913–24.

Beck, M. B. (1983) Sensitivity analysis, calibration and validation, in *Mathematical Modeling of Water Quality: Streams, Lakes and Reservoirs* (ed. G. T. Orlob), *International Series on Applied Systems Analysis*, Volume 12, Wiley, Chichester, pp. 425–67.

Beddington, J. R. (1974) Age distribution and the stability of simple discrete time population models. *Journal of Theoretical Biology*, **47**, 65–74.

Begon, M., Harper, J. L. and Townsend, C. R. (1986) *Ecology: Individuals, Populations and Communities*, Blackwell, Oxford.

Bellows, T. S. (1981) The descriptive properties of some models for density dependence. *Journal of Animal Ecology*, **50**, 139–56.

Bengtsson, K., Prentice, H. C., Rosen, E., Moberg, R. and Sjogren, E. (1988) The dry alvar grasslands of Oland: ecological amplitudes of plant species in relation to vegetation composition. *Acta Phytogeographica Sued*, **76**, 21–46.

Bernadelli, H. (1941) Population waves. *Journal of the Burma Research Society*, **31**, 1–18.

Bernstein, C. (1986) Density dependence and the stability of host-parasitoid systems. *Oikos*, **47**, 176–80.

Berry, R. J. (1971) Conservation aspects of the genetical constitution of populations, in *The Scientific Management of Animal and Plant Communities for Conservation* (eds E. Duffey and A. S. Watt), Blackwell, Oxford, pp. 177–206.

Besançon, F. (1984) *Contribution to the study of the biology and reproductive strategy of* Crocidura russula *in temperate environments*. PhD thesis, University of Lausanne, Lausanne (in French).

Beverton, R. J. H. and Holt, S. J. (1957) On the dynamics of exploited fish populations. (Great Britain) *Ministry of Agriculture, Fisheries and Food. Fishery Investigations* (series 2), **19**, 5–533.

Bierzychudek, P. (1982) The demography of Jack-in-the-pulpit, a forest perennial that changes sex. *Ecological Monographs*, **52**, 335–51.

Bleich, V. C., Wehausen, J. D. and Holl, S. A. (1990) Desert-dwelling mountain sheep: conservation implications of a naturally fragmented distribution. *Conservation Biology*, **4**, 383–90.

Boecklen, W. J. (1986) Optimal design of nature reserves: consequences of genetic drift. *Biological Conservation*, **38**, 323–38.

Boecklen, W. J. and Gotelli, N. J. (1984) Island biogeographic theory and conservation practice: Species-area or specious-area relationships? *Biological Conservation*, **29**, 63–80.

Bonnell, M. L. and Selander, R. K. (1974) Elephant seals: genetic variation and near extinction. *Science*, **184**, 908–9.

Borlase, S. C., Loebel, D. A., Frankham, R. *et al.* (1992) Modelling problems in conservation genetics using captive *Drosophila* populations: consequences of equalisation of family sizes. *Conservation Biology*, in press.

Boyce, M. S. (1977) Population growth with stochastic fluctuations in the life table. *Theoretical Population Biology*, **12**, 366–73.

Brewer, G. D. and deLeon, P. (1983) *The Foundations of Policy Analysis*, Dorsey Press, Homewood, Illinois.

Brillinger, D. R. (1986) The natural variability of vital rates and associated statistics. *Biometrics*, **42**, 693–734.

Briscoe, D. A., Malpica, J. M., Robertson, A. *et al.* (1992) Rapid loss of genetic variation in large captive populations of *Drosophila* flies: implications for the genetic management of captive populations. *Conservation Biology* (submitted).

Brotschol, J. V., Roberts, J. H. and Namkoong (1986) Allozyme variation among North Carolina populations of *Liriodendron tulipifera* L. *Silvae Genetica*, **35**, 131–8.

Brown, J. H. (1984) On the relationship between abundance and distribution of species. *American Naturalist*, **124**, 255–79.

Brown, A. H. D. and Weir, B. S. (1983) Measuring genetic variability in plant populations, in *Isozymes in plant genetics and breeding* (eds S. D. Tansley and T. J. Orton), Elsevier, Amsterdam, pp. 219–39.

Burbidge, A. A. and McKenzie, N. L. (1989) Patterns in the decline of the mammal fauna of Western Australia: causes and implications for future conservation. *Biological Conservation*, **50**, 143–98.

Burgman, M. A. (1989) The habitat volumes of scarce and ubiquitous plants: a test of the model of environmental control. *American Naturalist*, **133**, 228–39.

Burgman, M. A. and Gerard, V. A. (1990) A stage-structured, stochastic population model for the giant kelp *Macrocystis pyrifera*. *Marine Biology*, **105**, 15–23.

Burgman, M. A. and Lamont, B. B. (1992) A stochastic model for the viability of *Banksia cuneata* populations: environmental, demographic and genetic effects. *Journal of Applied Ecology*, **29**, 719–727.

Burgman, M. A. and Neet, C. R. (1989) Analyse des risques d'extinction des populations naturelles. *Acta Oecologia/Oecologia Generalis*, **10**, 233–43.

Burgman, M. A., Akçakaya, H. R. and Loew, S. S. (1988) The use of extinction models for species conservation. *Biological Conservation*, **43**, 9–25.

Burgman, M. A., Cantoni, D. and Vogel, P. (1992) Shrews in suburbia: an application of Goodman's extinction model. *Biological Conservation*, **61**, 117–123.

Burns, B. R. and Ogden, J. (1985) The demography of the temperate mangrove (*Avicennia marina* (Forsk.) Vierh.) at its southern limit in New Zealand. *Australian Journal of Ecology*, **10**, 125–33.

Cantoni, D. and Vogel, P. (1989) Social organisation and mating system of free-ranging greater white-toothed shrews, *Crocidura russula*. *Animal Behaviour*, **38**, 205–14.

Caputi, N. (1988) Factors affecting the time series bias in stock-recruitment relationships and the interaction between times series and measurement error bias. *Canadian Journal of Aquatic Science*, **45**, 178–84.

Carey, A. B., Reid, J. A. and Horton, S. P. (1990) Spotted owl home range and habitat use in southern Oregon Coast Ranges. *Journal of Wildlife Management*, **54**, 11–17.

Caswell, H. (1978) A general formula for the sensitivity of population growth rate to changes in life history parameters. *Theoretical Population Biology*, **14**, 215–30.

Caswell, H. (1986) Life cycle models for plants. *Lectures on Mathematics in the Life Sciences*, **18**, 171–233.

Caswell, H. (1989) *Matrix Population Models: Construction, Analysis and Interpretation*, Sinauer, Sunderland, Massachusetts.

Caughley, G. (1977) *Analysis of Vertebrate Populations*, Wiley, Chichester.

Chesser, R. K., Smith, M. H. and Brisbin, I. L. Jr (1980) Management and maintenance of genetic variability in endangered species. *International Zoo Yearbook*, **20**, 146–54.

Chesson, P. (1978) Predator–prey theory and variability. *Annual Review of Ecology and Systematics*, **9**, 323–47.

Chesson, P. L. and Warner, R. R. (1981) Environmental variability promotes coexistence in lottery competitive systems. *American Naturalist*, **117**, 923–43.

Clarke, B. (1972) Density-dependent selection. *American Naturalist*, **106**, 1–13.

Coates, D. J. (1988) Genetic diversity and population genetic structure in the rare Chittering grass wattle, *Acacia anomala* Court. *Australian Journal of Botany*, **36**, 273–86.

Cody, M. L. (1986) Diversity, rarity and conservation in Mediterranean regions, in *Conservation Biology: The Science of Scarcity and Diversity* (ed. M. E. Soulé), Sinauer, Sunderland, Massachusetts, pp. 122–52.

Cohen, J. E. (1979) Comparative static and stochastic dynamics of age-structured populations. *Theoretical Population Biology*, **16**, 159–71.

Collar, N. J. and Stuart, S. N. (1985) *Threatened Birds of Africa and Related Islands*. The ICBP, IUCN Red Data Book, Part 1 (3rd edn), ICBP, Cambridge.

Collins, N. M., Sayer, J. A. and Whitmore, T. C. (1991) *The Conservation Atlas of Tropical Forests: Asia and the Pacific*, Macmillan, New York.

Connell, J. H. (1983) On the prevalence and relative importance of interspecific competition: evidence from field experiments. *American Naturalist*, **122**, 661–96.

Conner, R. N. (1988) Wildlife populations: minimally viable or ecologically functional? *Wildlife Society Bulletin*, **16**, 80–4.

Conway, A. J. and Goodman, P. S. (1989) Population characteristics and management of black rhinoceros *Diceros bicornis minor* and white rhinoceros *Ceratotherium simum simum* in Ndumu Game Reserve, South Africa. *Biological Conservation*, **47**, 109–22.

Couvet, D., Gouyon, P., Kjellberg, F., Olivieri, I., Pomente, D. and Valdeyron, G. (1985) De la métapopulation an voisinage: la génétique des populations en déséquilibre. *Génétique Sélection Evolution*, **17**, 407–14.

Cowling, R. M., Lamont, B. B. and Enright, N. J. (1990) Fire and the management of south-western Australian banksias. *Proceedings of the Ecological Society of Australia*, **16**, 177–83.

Cronin, T. M. and Schneider, C. E. (1990) Climatic influences on species: evidence from the fossil record. *Trends in Ecology and Evolution*, **5**, 275–9.

Crouse, D. T., Crowder L. B. and Caswell, H. (1987) A stage-based population model for loggerhead sea turtles and implications for conservation. *Ecology*, **68**, 1412–23.

Crow, J. F. (1979) Gene frequency and fitness change in an age-structured population. *Annals of Human Genetics*, **42**, 355–70.

Crow, J. F. and Denniston, C. (1988) Inbreeding and variance effective population numbers. *Evolution*, **42**, 482–95.

Crow, J. F. and Kimura, M. (1970) *An Introduction to Population Genetics Theory*, Burgess, Minneapolis, Minnesota.

Crow, J. F. and Kimura, M. (1972) The effective number of a population with overlapping generations: a correction and further discussion. *American Journal of Human Genetics*, **24**, 1–10.

Crow, J. F. and Morton, N. E. (1955) Measurement of gene frequency drift in small populations. *Evolution*, **9**, 202–14.

Csetenyi, A. I. and Logofet, D. O. (1989) Leslie model revisited: some generalizations to block structures. *Ecological Modelling*, **48**, 277–90.

Darlington, C. D. (1958) *Evolution of Genetic Systems*, Oliver & Boyd, London.

Deevey, E. S. (1947) Life tables for natural populations of animals. *Quarterly Review of Biology*, **22**, 283–314.

de Kroon, H., Plaisier, A., van Groenendael, J. and Caswell, H. (1986) Elasticity: the relative contribution of demographic parameters to population growth rate. *Ecology*, **67**, 1427–31.

den Boer, P. J. (1968) Spreading of risk and stabilization of animal numbers. *Acta Biotheoretica*, **18**, 165–94.

den Boer, P. J. (1981) On the survival of populations in a heterogeneous and variable environment. *Oecologia*, **50**, 39–53.

Diamond, J. M. (1975) The island dilemma: lessons of modern biogeographic studies for the design of natural reserves. *Biological Conservation*, **7**, 129–46.

Diamond, J. M. (1984) 'Normal' extinctions of isolated populations, in *Extinctions* (ed. M. H. Nitecki), University of Chicago Press, Chicago, pp. 191–246.

Diamond, J. M. (1987) Extant unless proven extinct? Or, extinct unless proven extant. *Conservation Biology*, **1**, 77–9.

Dickman, C. (1986) Return of the phantom dibbler, *Australian Natural History*, **22**, 33.

Dickman, C. R. and Doncaster, C. P. (1989) The ecology of small mammals in urban habitats. H. Demography and dispersal. *Journal of Animal Ecology*, **58**, 119–27.

Dixon, K. R. and Juelson, T. C. (1987) The political economy of the spotted owl. *Ecology*, **68**, 772–6.

Dobson, A. (1988) *Some simple models for the dynamics of cracid populations*. Paper presented to the second International Symposium on the Biology and Conservation of the Family Carcidae, Caracas, February, 1988.

Dobson, A. P. and Lyles, A. M. (1989) The population dynamics and conservation of primate populations. *Conservation Biology*, **3**, 362–80.

Dobson, A., Jolly, A. and Rubenstein, D. (1989) The greenhouse effect and biological diversity. *Trends in Ecology and Evolution*, **4**, 64–8.

Drury, W. H. (1974) Rare species. *Biological Conservation*, **6**, 162–69.

Eberhardt, L. L. (1987) Population projections from simple models. *Journal of Applied Ecology*, **24**, 103–18.

Ehrlich, P. R. (1961) Intrinsic barriers to dispersal in the checkerspot butterfly. *Science*, **134**, 108–9.

Ehrlich, P. R. (1986) Extinction: What is happening now and what needs to be done, in *Dynamics of Extinction* (ed. D. K. Elliott), Wiley, New York, pp. 157–64.

Ehrlich, P. R. and Ehrlich, A. (1990) *The Population Explosion*, Simon and Schuster, New York.

Ehrlich, P. R., Breedlove, D. E., Brussard, P. F. and Sharp, M. A. (1972) Weather and the 'regulation' of subalpine populations. *Ecology*, **53**, 243–7.

Ehrlich, P. R., Murphy, D. D., Singer, M. C., Sherwood, C. B., White, R. R. and Brown, I. L. (1980) Extinction, reduction, stability and increase: the responses of checkerspot butterfly (*Euphydryas*) populations to the California drought. *Oecologia*, **46**, 101–5.

Ehrlich, P. R., White, R. R., Singer, M. C., McKechnie, S. W. and Gilbert, L. E. (1975) Checkerspot butterflies: a historical perspective. *Science*, **188**, 221–8.

Emlen, J. M. (1984) *Population Biology: The Coevolution of Population Dynamics and Behaviour*, Macmillan, New York.

Enright, N. and Ogden, J. (1970) Applications of transition matrix models in forest dynamics: *Araucana* in Papua, New Guinea and *Nothofagus* in New Zealand. *Australian Journal of Ecology*, **4**, 1–23.

Erwin, T. L. (1991) How many species are there?: revisited. *Conservation Biology*, **5**, 330–3.

Ewens, W. J. (1982) On the concept of effective population size. *Theoretical Population Biology*, **21**, 373–8.

Ewens, W. J., Brockwell, P. J., Gani, J. M. and Resnick, S. I. (1987) Minimum viable population size in the presence of catastrophes, in *Viable Populations for Conservation* (ed. M. E. Soulé), Cambridge University Press, Cambridge, pp. 59–68.

Faddeeva, V. N. (1959) *Computational Methods of Linear Algebra*, Dover Publications, New York.

Fahrig, L. and Merriam, G. (1985) Habitat patch connectivity and population survival. *Ecology*, **66**, 1762–8.

Falconer, D. S. (1989) *Introduction to Quantitative Genetics* (3rd edn), Longman, London.

Feller, W. (1939) Die Grundlagen der Volterrasche Theorie des Kampfes Wahrscheinlichkeitstheoretischer Behandlung. *Acta Biotheoretica*, **5**, 11–40.

Ferson, S. and Akçakaya, H. R. (1990) *RAMAS/age User Manual*, Applied Biomathematics, Setauket, New York.

Ferson, S. and Burgman, M. A. (1990) The dangers of being few: demographic risk analysis for rare species extinction, in *Ecosystem Management: Rare Species and Significant Habitats* (eds R. S. Mitchell, C. J. Sheviak and D. J. Leopold). *New York State Museum Bulletin*, **471**, 129–32.

Ferson, S., Downey, P., Klerks, P., Weissburg, M., Kroot, J., Stewart, S., Jacquez, G., Semakula, J., Malenky, R. and Anderson, K. (1986) Competing reviews or why do Connell and Schoener disagree? *American Naturalist*, **127**, 571–6.

Ferson, S., Ginzburg, L. and Silvers, A. (1989) Extreme event risk analysis for age-structured populations. *Ecological Modelling*, **47**, 175–87.

Fielder, P. L. and Jain, S. K. (1990) *Conservation Biology: The Theory and Practice of Nature Conservation, Preservation and Management*, Chapman and Hall, London.

Fisher, R. A. (1930) *The Genetical Theory of Natural Selection*, Clarendon Press, Oxford.

Flipse, E. and Veling, E. J. M. (1984) An application of the Leslie matrix model to the population dynamics of the hooded seal, *Cystophora cristata* Erxleben. *Ecological Modelling*, **24**, 43–59.

Fogarty, M. J., Sissenwine, M. P. and Cohen, E. B. (1991) Recruitment variability and the dynamics of exploited marine populations. *Trends in Ecology and Evolution*, **6**, 241–6.

Foose, T. J. and Ballou, J. D. (1988) Population management: theory and practice. *International Zoo Yearbook*, **27**, 26–41.

Forney, K. A. and Gilpin, M. E. (1989) Spatial structure and population extinction: a study with Drosophila flies. *Conservation Biology*, **3**, 45–51.

Frankel, O. H. and Soulé, M. E. (1981) *Conservation and Evolution*, Cambridge University Press, Cambridge.

Franklin, I. R. (1980) Evolutionary change in small populations, in *Conservation Biology: An Evolutionary-Ecological Perspective* (eds M. Soulé and B. Wilcox), Sinauer, Massachusetts, pp. 135–49.

Fritz, R. S. (1979) Consequences of insular population structure: distribution and extinction of spruce grouse populations. *Oecologia*, **42**, 57–65.

Fulda, J. S. (1981) The logistic equation and population decline. *Journal of Theoretical Biology*, **91**, 255–9.

Futuyma, D. (1986) *Evolutionary Biology* (2nd edn), Sinauer, Sunderland, Massachusetts.

Garrett, T. (1972) *Report on the International Whaling Commission Meeting*. Information Report, Animal Welfare Institute, Washington, DC.

Gaston, K. J. (1991) The magnitude of global insect species richness. *Conservation Biology*, **5**, 283–96.

Gaston, K. J. and Lawton, J. H. (1987) A test of statistical techniques for detecting density dependence in sequential censuses of animal populations. *Oecologia*, **74**, 404–10.

Geng, S., Pennig de Vries, F. and Supit, I. (1986) A simple method for generating daily rainfall data. *Agricultural and Forest Meteorology*, **36**, 363–76.

Genoud, M. (1981) *Contribution to the study of the energetic strategy and ecological distribution of* Crocidura russula *in temperate environments*. PhD thesis, University of Lausanne, Lausanne (in French).

Genoud, M. and Hausser, J. (1979) Ecology of a population of *Crocidura russula* in a rural montaine environment. *Terre et la Vie*, **33**, 539–54 (in French).

Getz, W. M. and Haight, R. G. (1989) *Population Harvesting: Demographic Models of Fish, Forest, and Animal Resources*, Princeton University Press, Princeton.

Gilbert, N. and Lee, S. B. (1980) Two perils of plant population dynamics. *Oecologia*, **46**, 283–4.

Gill, D. W. (1978) The metapopulation ecology of the red-spotted newt, *Notophthalmus viridescens* (Rafinesque). *Ecological Monographs*, **48**, 145–66.

Gilpin, M. E. (1988) A comment on Quinn and Hastings: extinction in subdivided habitats. *Conservation Biology*, **2**, 290–2.

Gilpin, M. E. (1990) Extinction of finite metapopulations in correlated environments. In *Life in a Patchy Environment* (ed. Shoorocks), Oxford University Press.

Gilpin, M. (1991) The genetic effective size of a metapopulation. *Biological Journal of the Linnean Society*, **42**, 165–75.

Gilpin, M. E. and Diamond, J. M. (1976) Calculation of immigration and extinction curves from the species-area-distance relation. *Proceedings of the National Academy of Science, USA*, **73**, 4130–4.

Gilpin, M. E. and Soulé, M. E. (1986) Minimum viable populations: processes of species extinctions, in *Conservation Biology: the Science of Scarcity and Diversity* (ed. M. E. Soulé), Sinauer, Sunderland, Massachusetts, pp. 19–34.

Gingerich, P. D. (1985) Species in the fossil record: concepts, trends and transitions. *Paleobiology*, **11**, 27–41.

Ginzburg, L. R. (1984) *Assessment of Ecological Risk*. Research for managing the nation's estuaries, UNC Sea Grant Publication 84-08, pp. 1–21.

Ginzburg, L. R. and Pugliese, A. (1983) Extinction probabilities in stochastic age-structured models of population growth. Proceedings of the International Conference on Population Biology. Edmonton, Canada. *Lecture Notes in Biomathematics*, **52**, 154–62.

Ginzburg, L. R., Ferson, S. and Akçakaya, H. R. (1990) Reconstructibility of density dependence and the conservative assessment of extinction risks. *Conservation Biology*, **4**, 63–70.

Ginzburg, L. R., Johnson, K., Pugliese, A. and Gladden, J. (1984) Ecological risk assessment based on stochastic age-structured models of population growth. *Special Technical Testing Publication*, **845**, 31–45.

Ginzburg, L. R., Slobodkin, L. B., Johnson, K. and Bindman, A. G. (1982) Quasiextinction probabilities as a measure of impact on population growth. *Risk Analysis*, **21**, 171–81.

Given, D. (1990) Conserving botanical diversity on a global scale. *Annals of the Missouri Botanical Gardens*, **77**, 48–62.

Gliwicz, J. (1988) The role of dispersal in models of small rodent population dynamics. *Oikos*, **52**, 219–21.

Goodman, D. (1980) Demographic intervention for closely managed populations, in *Conservation Biology An evolutionary-ecological perspective*, (eds M. E. Soulé and B. A. Wilcox), Sinauer, Sunderland, Massachusetts, pp. 171–95.

Goodman, D. (1982) Optimal life histories, optimal notation, and the value of reproductive value. *American Naturalist*, **119**, 803–23.

Goodman, D. (1984) Risk spreading as an adaptive strategy in iteroparous life histories. *Theoretical Population Biology*, **25**, 1–20.

Goodman, D. (1987a) Considerations of stochastic demography in the design and management of biological reserves. *Natural Resources Modeling*, **1**, 205–34.

Goodman, D. (1987b) The demography of chance extinction, in *Viable Populations for Conservation* (ed. M. E. Soulé), Cambridge University Press, Cambridge, pp. 11–34.

Goodman, D. (1987c) How do any species persist? Lessons for conservation biology. *Conservation Biology*, **1**, 59–62.

Goodman, L. A. (1969) The analysis of population growth when birth and death rates depend upon several factors. *Biometrics*, **25**, 659–81.

Goodnight, C. J. (1988) Epistasis and the effect of founder events on the additive genetic variance. *Evolution*, **42**, 441–54.

Gould, G. I. (1985) Management of spotted owls by the California Department of Fish and Game, in *Ecology and Management of the Spotted Owl in the Pacific Northwest* (eds R. J. Gutierrez and A. B. Carey), U.S. Department of Agriculture, Forest Service, Pacific Northwest Forest and Range Experiment Station, General Technical Report PNW-185, pp. 21–26.

Gould, S. J. and Eldredge, N. (1977) Punctuated equilibria: the tempo and mode of evolution reconsidered. *Paleobiology* **3**, 115–151.

Greig, J. C. (1979) Principles of genetic conservation in relation to wildlife management in southern Africa. *South African Journal of Wildlife Research*, **9**, 57–78.

Griffith, B., Scott, J. M., Carpenter, J. W. and Reed, C. (1989) Translocation as a species conservation tool: status and strategy. *Science*, **245**, 477–80.

Gumbel, E. J. (1958) *Statistics of Extremes*, Columbia University Press, New York.

Gutierrez, R. J. and Carey, A. B. (eds) (1985) *Ecology and Management of the Spotted Owl in the Pacific Northwest*, U.S.D.A., Forest Service, Pacific Northwest Forest and Range Experiment Station, Portland, Oregon. General Technical Report PNW-185.

Gyllenberg, M. and Hanski, I. (1990) *Single-Species Metapopulation Dynamics: A Structured Model*, Lulea University of Technology, Department of Applied Mathematics, Research Report 11.

Hall, N., Wainwright, R. W. and Wolf, L. J. (1981) *Summary of Meteorological Data in Australia*, CSIRO Division of Forest Research, Report Number **6**, Canberra.

Hallam, T. G. and Clark, C. E. (1981) Non-autonomous equations as models of populations in a deteriorating environment. *Journal of Theoretical Biology*, **93**, 303–11.

Hammarling, S. J. (1970) *Latent Roots and Latent Vectors*, University of Toronto Press.

Hamrick, J. L. (1983) The distribution of genetic variation within the among natural plant populations, in *Genetics and Conservation: A Reference for Managing Wild Animal and Plant Populations* (eds C. M. Schonewald-Cox, S. M. Chambers, B. MacBryde and L. Thomas), Benjamin-Cummings, Menlo Park, California, pp. 335–48.

Hamrick, J. L., Linhart, Y. B. and Mitton, J. B. (1979) Relationship between life history characteristics and electrophoretically detectable genetic variation in plants. *Annual Review of Ecology and Systematics*, **10**, 173–200.

Hanski, I. (1983) Coexistence of competitors in patchy environment. *Ecology*, **64**, 493–500.

Hanski, I. (1989) Metapopulation dynamics: does it help to have more of the same? *Trends in Ecology and Evolution*, **4**, 113–14.

Hanski, I. (1991) Single-species metapopulation dynamics: concepts, models and observations. *Biological Journal of the Linnean Society*, **42**, 17–38.

Hanski, I. and Gilpin, M. (1991) Metapopulation dynamics: brief history and conceptual domain. *Biological Journal of the Linnean Society*, **42**, 3–16.

Hansson, L. (1991) Dispersal and connectivity in metapopulations. *Biological Journal of the Linnean Society*, **42**, 89–103.

Harcourt, A. H., Fossey, D. and Sabater-Pi, J. (1981) Demography of Gorilla gorilla. *Journal of Zoology*, **195**, 215–33.

Harper, J. L. (1981) The meanings of rarity, in *The Biological Aspects of Rare Plant Conservation* (ed. H. Synge), Wiley, Sydney, pp. 189–203.

Harrison, S. (1991) Local extinction in a metapopulation context: an empirical evaluation. *Biological Journal of the Linnean Society*, **42**, 73–88.

Harrison, S. and Quinn, J. F. (1989) Correlated environments and the persistence of metapopulations. *Oikos*, **56**, 293–8.

Harrison, S., Murphy, D. D. and Ehrlich, P. R. (1988) Distribution of the bay checkerspot butterfly *Euphydryas editha bayensis*: evidence for a metapopulation model. *American Naturalist*, **132**, 360–82.

Hartshorn, G. S. (1975) A matrix model of tree population dynamics, in *Tropical Ecological Systems* (eds F. B. Golley and E. Medina), Springer-Verlag, New York.

Hassell, M. P. (1986) Detecting density dependence. *Trends in Ecology and Evolution*, **1**, 90–3.

Hassell, M. P. and Sabelis, M. W. (1987) Reply. *Trends in Ecology and Evolution*, **2**, 78.

Hassell, M. P., Latto, J. and May, R. M. (1989) Seeing the wood for the trees: detecting density dependence from existing life-table studies. *Journal of Animal Ecology*, **58**, 883–92.

Hassell, M. P., Lawton, J. H. and May, R. M. (1976) Patterns of dynamical behaviour in single-species populations. *Journal of Animal Ecology*, **45**, 471–86.

Hassell, M. P., Southwood, T. R. E. and Reader, P. M. (1987) The dynamics of the viburnum whitefly (*Aleurotrachelus jelinekii*): a case study of population regulation. *Journal of Animal Ecology*, **56**, 283–300.

Hastings, A. and Wolin, C. L. (1989) Within patch dynamics in a metapopulation. *Ecology*, **70**, 1261–6.

Henderson, M. T., Merriam, G. and Wegner, J. (1985) Patchy environments and species survival: chipmunks in an agricultural mosaic. *Biological Conservation*, **31**, 95–1005.

Hilborn, R. (1987) Living with uncertainty in resource management. *North American Journal of Fisheries Management*, **7**, 1–5.

Hill, W. G. (1979) A note on effective population size with overlapping generations. *Genetics*, **92**, 317–22.

Hobbs, R. J. and Hopkins, A. J. M. (1990) From frontier to fragments: European impact on Australia's vegetation. *Proceedings of the Ecological Society of Australia*, **16**, 93–114.

Holling, C. S. (1973) Resilience and stability of ecological systems. *Annual Review of Ecology and Systematics*, **4**, 1–23.

Holling, C. S. (1978) *Adaptive Environmental Assessment and Management*, Wiley, Chichester.

Hopper, S. D. (1979) Biogeographical aspects of speciation in the Southwest Australian flora. *Annual Review of Ecology and Systematics*, **10**, 399–422.

Hopper, S. D. and Coates, D. J. (1990) Conservation of genetic resources in Australia's flora and fauna. *Proceedings of the Ecological Society of Australia*, **16**, 567–77.

Houllier, F. and Lebreton J.-D. (1986) A renewal equation approach to the dynamics of stage-grouped populations. *Mathematical Biosciences*, **79**, 185–97.

Houllier, F., Lebreton, J.-D, and Pontier, D. (1989) Sampling properties of the asymptotic behaviour of age- or stage-grouped population models. *Mathematical Biosciences*, **95**, 161–77.

Hubbell, S. B. and Werner, P. A. (1979) On measuring the intrinsic rate of increase of populations with heterogeneous life histories. *The American Naturalist*, **113**, 277–93.

Huenneke, L. F. and Marks, P. L. (1987) Stem dynamics of the shrub *Alnus incana* ssp. *rugosa*: transition matrix models. *Ecology*, **68**, 1234–42.

Hughes, T. P. (1984) Population dynamics based on individual size rather than age: a general model with a reef coral example. *The American Naturalist*, **123**, 778–95.

Hughes, T. P. and Connell, J. H. (1987) Population dynamics based on size or age? A reef coral analysis. *The American Naturalist*, **129**, 818–29.

Imberger, J. (1985) Thermal characteristics of standing waters: an illustration of dynamic processes. *Hydrobiologia*, **125**, 7–29.

Jablonski, D. (1986) Background and mass extinctions: the alternation of macroevolutionary regimes. *Science*, **231**, 129–33.

James, S. H. (1982) The relevance of genetic systems in *Isotoma petraea* to conservation practice, in *Species at Risk: Research in Australia* (eds R. H. Groves and W. D. L. Ride), Australian Academy of Science, Canberra, pp. 63–71.

Janzen, D. H. (1976) Why bamboos wait so long to flower. *Annual Review of Ecology and Systematics*, **7**, 347–91.

Janzen, D. H. (1988) Tropical, ecological and biocultural restoration. *Science*, **239**, 243–4.

Jenkins, M. B. (1987) *Madagascar: An Environmental Profile*, IUCN, Gland, Switzerland.

Jenkins, S. H. (1988) Use and abuse of demographic models of population growth. *Bulletin of the Ecological Society of America*, **69**, 201–7.

Jordan, W. R. jr, Gilpin, M. E. and Aber, J. D. (1990) *Restoration Ecology*, Cambridge University Press, Cambridge.

Kemeny, J. G. and Kurtz, T. E. (1985) *True BASIC: The Structural Language System for the Future*, Addison-Wesley, Reading, Massachusetts.

Kendall, D. G. (1948) On the generalised birth and death process. *Annals of Mathematics and Statistics*, **19**, 1–15.

Kenward, R. E. and Holm, J. L. (1989) What future for British red squirrels? *Biological Journal of the Linnean Society*, **38**, 83–9.

Keyfitz, N. (1968) *Introduction to the Mathematics of Populations*, Addison-Wesley, Reading, Massachusetts.

Keyfitz, N. (1972) On future population. *Journal of the American Statistical Association*, **67**, 347–63.

Kimura, M. and Crow, J. F. (1963) The measurement of effective population number. *Evolution*, **17**, 279–88.

Kirkpatrick, M. (1984) Demographic models based on size, not age, for organisms with indeterminate growth. *Ecology*, **65**, 1874–84.

Kitching, R. (1971) A simple simulation model of dispersal of animals among units of discrete habitat. *Oecologia*, **7**, 95–116.

Knuth, D. E. (1981) *The Art of Computer Programming, Volume 2: Seminumerical Algorithms* (2nd edn), Addison-Wesley, Reading, Massachusetts.

Koenig, W. D. (1988) On determination of viable population size in birds and mammals. *Wildlife Society Bulletin*, **16**, 230–4.

Krebs, C. J. (1985) *Ecology: The Experimental Analysis of Distribution and Abundance*, Harper & Row, New York.

Kuno, E. (1987) Principles of predator-prey interaction in theoretical, experimental, and natural population systems. *Advances in Ecological Research*, **16**, 250–337.

Lamont, B. B. and Barker, M. J. (1988) Seed bank dynamics of a serotinous fire-sensitive *Banksia* species. *Australian Journal of Botany*, **36**, 193–203.

Lamont, B. B., Connell, S. W. and Bergl, S. M. (1991) Seed bank and population dynamics of *Banksia cuneata*: the role of time, fire, and moisture, *Botanical Gazette*, **152**, 114–22.

Lande, R. (1981) The minimum number of genes contributing to quantitative variation between and within populations, *Genetics*, **99**, 541–53.

Lande, R. (1988a) Demographic models of the northern spotted owl *(Strix occidentalis caurina)*. *Oecologia*, **75**, 601–7.

Lande, R. (1988b) Genetics and demography in biological conservation. *Science*, **241**, 1455–60.

Lande, R. and Barrowclough, G. F. (1987) Effective population size, genetic variation, and their use in population management, in *Viable Populations for Conservation* (ed. M. E. Soulé), Cambridge University Press, Cambridge, pp. 87–123.

Lande, R. and Orzack, S. H. (1988) Extinction dynamics of age-structured populations in a fluctuating environment. *Proceedings of the National Academy of Science, USA*, **85**, 7418–21.

Law, R. (1983) A model for the dynamics of a plant population containing individuals classified by age and size. *Ecology*, **64**, 224–30.

Law, R. and Edley, M. T. (1990) Transient dynamics of populations with age- and stage dependent vital rates. *Ecology*, **71**, 1863–70.

Lebreton, J. D. (1978) A probabilistic model of the population dynamics of the white swan (*Ciconia ciconia* L.) in Western Europe, in *Biometry and Ecology* (eds J. M. Legay and P. Tomassone), Société Française de Biometrie (in French), pp. 277–343.

Lefkovitch, L. P. (1965) The study of population growth in organisms grouped by stages. *Biometrics*, **21**, 1–18.

Lefkovich, L. P. and Fahrig, L. (1985) Spatial characteristics of habitat patches and population survival. *Ecological Modelling*, **30**, 297–308.

Lehman, R. S. (1977) *Computer Simulation and Modeling: An Introduction*, Wiley, New York.

Lehmkuhl, J. F. (1984) Determining size and dispersion of minimum viable populations for land management planning and species conservation. *Environmental Management*, **8**, 167–76.

Leigh, E. G. Jr (1981) The average lifetime of a population in a varying environment. *Journal of Theoretical Biology*, **90**, 213–39.

Leigh, J. H., Briggs, J. D. and Hartley, W. (1981) Rare or threatened Australian plants. *Australian National Parks and Wildlife Service Special Publication*, **7**, 1–178.

Leslie, P. H. (1945) On the use of matrices in certain population mathematics. *Biometrika*, **33**, 183–212.

Leslie, P. H. (1948) Some further notes on the use of matrices in population mathematics. *Biometrika*, **35**, 213–45.

Leslie, P. H. (1959) The properties of a certain lag type of population growth and the influence of an external, random factor on the number of such species. *Physiological Zoology*, **32**, 151–9.

Levins, R. (1966) The strategy of model building in population biology. *American Scientist*, **54**, 421–31.

Levins, R. (1969) The effect of random variation of different types on population growth. *Proceedings of the National Academy of Science, USA*, **62**, 1061–5.

Levins, R. (1970) Extinction, in *Some Mathematical Questions in Biology* (ed. M. Gerstenhaber), American Mathematical Society, Providence, Rhode Island.

Levinton, J. S. and Ginzburg, L. (1984) Repeatability of taxon longevity in successive foraminifera radiations and a theory of random appearance and extinction. *Proceedings of the National Academy of Science, USA*, **81**, 5478–81.

Lewis, E. G. (1942) On the generation and growth of a population. *Sankya*, **6**, 93–6.

Lewis, W. H. and Zenger, V. E. (1982) Population dynamics of the American ginseng *Panax quinquefolium* (Araliocere). *American Journal of Botany*, **69**, 1483–90.

Lewontin, R. C. (1974) *The Genetic Basis of Evolutionary Change*, Columbia University Press, New York.

Lewontin, R. C. and Cohen, D. (1969) On population growth in a randomly varying environment. *Proceedings of the National Academy of Science, USA*, **62**, 1056–60.

Lindenmayer, D. B., Clark, T. W., Lacy, R. C. and Thomas, V. C. (1992) Population viability analysis as a tool in wildlife conservation policy: with reference to Australia, *Environmental Management*, **17**, 745–758.

Lomnicki, A. (1987) Density vague ecology. *Trends in Ecology and Evolution*, **2**, 76.

Lomnicki, A. (1988) *Population Ecology of Individuals*, Princeton University Press, Princeton, New Jersey.

Lotka, A. J. (1924) *Elements of Physical Biology*, Williams & Wilkins, Baltimore. [republished in 1956 as *Elements of Mathematical Biology* by Dover Publications, New York].

Lovejoy, T. E., Bierregaard, R. O., Rylands, A. B. *et al.* (1986) Edge and other effects of isolation on Amazon forest fragments, in *Conservation, Biology: The Science of Scarcity and Diversity* (ed. M. E. Soulé), Sinauer, Sunderland, Massachusetts, pp. 257–285.

Loveless, M. D. and Hamrick, J. L. (1984) Ecological determinants of genetic structure in plant populations. *Annual Review of Ecology and Systematics*, **15**, 65–95.

Lyell, C. (1832) *Principles of Geology*, Volume 2, Murray, London.

MacArthur, R. H. (1972) *Geographical Ecology*, Harper & Row, New York.

MacArthur, R. H. and Wilson, E. O. (1967) *The Theory of Island Biogeography*, Princeton University Press, Princeton, New Jersey.

McCullough, D. R. (1979) *The George Reserve Deer Herd: Population Ecology of a K-selected Species*, The University of Michigan Press, Ann Arbor.

Mace, G. M. and Lande, R. (1991) Assessing extinction threats: towards a re-evaluation of IUCN threatened species categories. *Conservation Biology*, **5**, 148–57.

Maillette, L. (1982) Structural dynamics of silver birch: II. A matrix model of the bud population. *Journal of Applied Ecology*, **19**, 219–38.

Marcot, B. G., Carrier, D. and Holthausen, R. (1986) The northern spotted owl (*Strix occidentalis caunna*), in *The Management of Viable Populations: Theory, Applications, and Case Studies* (eds B. A. Wilcox, P. F. Brussard and B. G. Marcot), Center for Conservation Biology, Stanford University.

Mares, M. A. (1986) Conservation in South America: problems, consequences and solutions. *Science*, **233**, 734–9.

Margules, C. and Usher, M. B. (1981) Criteria used in assessing wildlife conservation potential: A review. *Biological Conservation*, **21**, 79–109.

Marshall, D. R. and Brown, A. H. D. (1975) Optimal sampling strategies in genetic conservation, in *Isozymes in Plant Genetic Resources for Today and Tomorrow* (eds O. H. Frankel and J. E. Hawkes), Cambridge University Press, Cambridge, pp. 53–80.

Marshall, L. G. (1988) Extinction, in *Analytical Biogeography* (eds A. A. Myers and P. S. Giller), Chapman and Hall, London, pp. 219–254.

Maruyama, T. (1972) Rate of decrease of genetic variability in a two-dimensional continuous population of finite size. *Genetics*, **70**, 639–51.

Maruyama, T. (1977) Stochastic problems in population genetics. *Lecture Notes in Biomathematics*, Volume 17, Springer-Verlag, Berlin.

May, R. M. (1973a) *Stability and Complexity in Model Ecosystems*, Princeton University Press, Princeton, New Jersey.

May, R. M. (1973b) Stability in randomly fluctuating versus deterministic environments. *American Naturalist*, **107**, 621–50.

May, R. M. (1976a) Models for single populations, in *Ecology: Principles and Applications* (ed. R. M. May), Blackwell, Oxford, pp. 4–25.

May, R. M. (1976b) *Theoretical Ecology*, Blackwell, Oxford.

May, R. M. (1988) How many species are there on earth? *Science*, **241**, 1441–9.

May, R. M. (1989) Detecting density dependence in imaginary worlds. *Nature*, **338**, 16–17.

May, R. M. (1990) How many species? *Philosophical Transactions of the Royal Society of London B*, **330**, 293–304.

May, R. M. and Oster, G. F. (1976) Bifurcations and dynamic complexity in simple ecological models. *American Naturalist*, **110**, 573–600.

May, R. M., Conway, G. R., Hassell, M. P. and Southwood, T. R. E. (1974) Time delays, density-dependence and single-species oscillations. *Journal of Animal Ecology*, **43**, 747–70.

Maynard-Smith, J. (1968) *Mathematical Ideas in Biology*, Cambridge University Press, Cambridge.

Maynard-Smith, J. (1974) *Models in Ecology*, Cambridge University Press, Cambridge.

Maynard-Smith, J. (1989) *Evolutionary Genetics*, Oxford University Press, Oxford.

Maynard-Smith, J. and Slatkin, M. (1973) The stability of predator-prey systems. *Ecology*, **54**, 384–91.

Mehrhoff, L. A. (1989) The dynamics of declining populations of an endangered orchid, *Isotria medeoloides*. *Ecology*, **70**, 783–6.

Menges, E. S. (1990) Population viability analysis for an endangered plant. *Conservation Biology*, **4**, 52–62.

Menges, E. S. and Gawler, S. C. (1986) Four-year changes in population size of the endemic Furbish's lousewort: implications for endangerment and management. *Natural Areas Journal*, **6**, 6–17.

Merriam, G. (1991) Corridors and connectivity: animal populations in heterogeneous environments, in *Nature Conservation 2: The Role of Corridors* (eds D. A. Saunders and R. J. Hobbs), Surrey Beatty, Chipping Norton, New South Wales, pp. 133–42.

Metz, J. A. J. and Diekmann, O. (1986) *The Dynamics of Physiologically Structured Populations*, Springer-Verlag, New York.

Middleton, J. and Merriam, G. (1981) Woodland mice in farmland mosaic. *Journal of Applied Ecology*, **18**, 703–10.

Middleton, J. and Merriam, G. (1983) Distribution of woodland species in farmland woods, *Journal of Applied Ecology*, **20**, 625–44.

Moloney, K. A. (1986) A generalised algorithm for determining category size. *Oecologia*, **69**, 176–80.

Moonan, W. J. (1957) Linear transformation to a set of stochastically dependent normal variables. *Journal of the American Statistical Association, USA*, **52**, 247–52.

Moore, N. W. (1962) The heaths of Dorset and their conservation. *Journal of Ecology*, **50**, 369–91.

Moran, G. F. and Hopper, S. D. (1983) Genetic diversity and the insular population structure of the rare granite rock species, *Eucalyptus caesia* Benth. *Australian Journal of Botany*, **31**, 161–72.

Moran, G. F. and Hopper, S. D. (1987) Conservation of genetic resources of rare and widespread eucalypts in remnant vegetation, in *Nature Conservation: The Role of Remnants of Vegetation* (eds D. A. Saunders, G. W. Arnold, A. A. Burbidge and A. J. M. Hopkins), Surrey Beatty, CSIRO, Western Australian Department of Conservation and Land Management, pp. 151–62.

Morris, W. F. (1990) Problems in detecting chaotic behaviour in natural populations by fitting simple discrete models. *Ecology*, **71**, 1849–62.

Mountford, M. D. (1988) Population regulation, density dependence, and heterogeneity. *Journal of Animal Ecology*, **57**, 845–58.

Munton, P. (1987) Concepts of threat and status of wild populations, in *The Road to Extinction* (eds R. Fitter and M. Fitter), International Union for the Conservation of Nature, Gland, Switzerland, pp. 72–95.

Murphy, D. D., Freas, K. E. and Weiss, S. B. (1990) An environment-metapopulation approach to population viability analysis for a threatened vertebrate. *Conservation Biology*, **4**, 41–51.

Myers, N. (1981) Conservation needs and opportunities in tropical moist forests, in *The Biological Aspects of Rare Plant Conservation* (ed. H. Synge), Wiley, New York, pp. 141–54.

Myers, N. (1988) Threatened biotas: 'hot spots' in tropical forests. *The Environmentalist*, **8**, 1–20.

Nectoux, F. and Kuroda, Y. (1989) *Timber From the South Seas. An Analysis of Japan's Tropical Timber Trade and its Environmental Impact*, World Wide Fund for Nature International.

Nei, M. (1973) Analysis of gene diversity in subdivided populations. *Proceedings of the National Academy of Science, USA*, **70**, 3321–3.

Nei, M. (1975) *Molecular Population Genetics and Evolution*, North Holland, Amsterdam.

Nei, M. (1978) Estimation of average heterozygosity and genetic distance from a small number of individuals. *Genetics*, **89**, 583–90.

Nei, M. and Graur, D. (1984) Extent of protein polymorphism and the neutral mutation theory. *Evolutionary Biology*, **17**, 73–118.

Nei, M., Chakravarti, A. and Tateno, Y. (1977) Mean and variance of FST in a finite number of incompletely isolated populations. *Theoretical Population Biology*, **11**, 291–306.

Nicholson, A. J. (1954) An outline of the dynamics of animal populations. *Australian Journal of Zoology*, **2**, 9–65.

Nisbet, R. M. and Gurney, W. S. C. (1982) *Modelling Fluctuating Populations*, Wiley, Chichester.

Nitecki, M. H. (1984) *Extinctions*, University of Chicago Press, Chicago.

Noon, B. R. and Biles, C. M. (1990) Mathematical demography of spotted owls in the Pacific northwest. *Journal of Wildlife Management*, **54**, 18–27.

O'Brien, S. J. (1989) Reply. *Trends in Ecology and Evolution*, **4**, 178.

O'Brien, S. J., Wildt, D. E. and Bush M. (1986) The cheetah in genetic peril. *Scientific American*, **254**, 68–76.

O'Neill, R. V., Gardner, R. and Mankin, J. B. (1980) Analysis of parameter error in a non-linear model. *Ecological Modelling*, **8**, 297–311.

Parsons, P. A. (1989) Environmental stresses and conservation of natural populations. *Annual Review of Ecology and Systematics*, **20**, 29–49.

Paton, G. P. (1986) A matrix modelling approach to population growth systems involving multiple time delays. *Ecological Modelling*, **34**, 197–216.

Pearl, R. and Miner, J. R. (1935) Experimental studies on the duration of life. XIV. The comparative mortality of certain lower organisms. *Quarterly Review of Biology*, **10**, 60–79.

Pennycuick, L. (1969) A computer model of the Oxford great tit population. *Journal of Theoretical Biology*, **22**, 381–400.

Pennycuick, L., Compton, R. and Beckingham, L. (1968) A computer model for simulating the growth of a population or of two interacting populations. *Journal of Theoretical Biology*, **18**, 316–29.

Peterson, C. H. and Black, R. (1988) Density-dependent mortality caused by physical stress interacting with biotic history. *American Naturalist*, **131**, 257–70.

Pielou, E. C. (1977) *Mathematical Ecology*, Wiley, New York.

Pimm, S. L., Gittleman, J. L., McCracken, G. F. and Gilpin, M. (1989) Plausible alternatives to bottlenecks to explain reduced genetic diversity. *Trends in Ecology and Evolution*, **4**, 176–8.

Pimm, S. L., Jones, H. L. and Diamond, J. (1988) On the risk of extinction. *American Naturalist*, **132**, 757–85.

Pinero, D., Martinez-Ramos, M. and Sarukhan, J. (1984) A population model of *Astrocaryum mexicanum* and a sensitivity analysis of its finite rate of increase. *Journal of Ecology*, **72**, 977–91.

Pokki, J. (1981) Distribution, demography and dispersal of the field vole *Microtus agrestis* (L.) in the Tvarminne archipelago, Finland. *Acta Zoologica Fennica*, **164**, 1–48.

Pollard, E. (1981) Resource limited and equilibrium models of predation. *Oecologia*, **49**, 277–8.

Pollard, E., Lakhani, K. H. and Rothery, P. (1987) The detection of density dependence from a series of annual censuses. *Ecology*, **68**, 2046–55.

Pollard, J. H. (1966) On the use of the direct matrix product in analyzing certain stochastic population models. *Biometrika*, **53**, 397–415.

Poole, R. W. (1974) *An Introduction to Quantitative Ecology*, McGraw-Hill, New York.

Possingham, H. and Noble, I. (1991) *An evaluation of population viability analysis for assessing the risk of extinction: a new model and its application to two species of forest fauna*. Report for the Forest and Timber Enquiry, Resource Assessment Commission, Canberra.

Possingham, H., Davies, I., Noble, I. R. and Norton, T. W. (1991a) A metapopulation simulation model for assessing the likelihood of plant and animal extinctions. *Proceedings of the 9th Biennial Conference, Simulation Society of Australia, Gold Coast, Queensland*.

Possingham, H., Noble, I. R. and Davies, I. (1991b) *ALEX: Analysis of the Likelihood of Extinction*, Department of Applied Mathematics, The University of Adelaide, South Australia.

Prentice, H. C. (1984) Enzyme polymorphism, morphometric variation and population structure in a restricted endemic, *Silene diclinis* (Caryophyllaceae). *Biological Journal of the Linnean Society*, **22**, 125–43.

Press, W. H., Flannery, B. P., Teukolsky, S. A. and Vetterling, W. T. (1986) *Numerical Recipes*, Cambridge University Press, Cambridge.

Prober, S. M. and Austin, M. P. (1991) Habitat peculiarity as a cause of rarity in *Eucalyptus paliformis*. *Australian Journal of Ecology*, **16**, 189–205.

Quinn, J. F. and Hastings, A. (1987) Extinction in subdivided habitats. *Conservation Biology*, **1**, 198–208.

Rabinowitz, D. (1978) Abundance and diaspore weight in rare and common prairie grasses. *Oecologia*, **37**, 213–19.

Rabinowitz, D., Cairns, S. and Dillon, T. (1986) Seven forms of rarity and their frequency in the flora of the British Isles, in *Conservation Biology: The Science of Scarcity and Diversity* (ed. M. E. Soulé), Sinauer, Sunderland, Massachusetts, pp. 182–204.

Ralls, K. and Ballou, J. (1982a) Effect of inbreeding on juvenile mortality in some small mammal species. *Laboratory Animals*, **16**, 159–66.

Ralls, K. and Ballou, J. (1982b) Effects of inbreeding on infant mortality in captive primates. *International Journal of Primatology*, **3**, 491–505.

Ralls, K. and Ballou, J. (1983) Extinction: lessons from zoos, in *Genetics and Conservation: A Reference for Managing Wild Animal and Plant Populations* (eds C. M. Schonewald-Cox, S. M. Chambers, B. MacBryde and L. Thomas), Benjamin/Cummings, Menlo Park, California, pp. 164–84.

Ralls, K., Ballou, J. and Brownell, R. L. (1983) Genetic diversity in Californian sea otters: Theoretical considerations and management implications. *Biological Conservation*, **25**, 209–32.

Ralls, K. Harvey, P. H. and Lyles, A. M. (1986) Inbreeding in natural populations of birds and mammals, in *Conservation Biology: The Science of Scarcity and Diversity* (ed. M. E. Soulé), Sinauer, Sunderland, Massachusetts, pp. 35–56.

Raup, D. M. and Gould, S. J. (1974) Stochastic simulation and evolution of morphology – towards a nomothetic paleontology. *Systematic Zoology*, **23**, 305–22.

Raup, D. M. and Sepkoski, J. J. (1982) Mass extinctions in the marine fossil record. *Science*, **215**, 1501–3.

Raup, D. M. and Sepkoski, J. J. (1984) Periodicity of extinctions in the geologic past. *Proceedings of the National Academy of Science, USA*, **81**, 801–5.

Recher, H. F. and Lim, L. (1990) A review of current ideas of the extinction, conservation and management of Australia's terrestrial vertebrate fauna. *Proceedings of the Ecological Society of Australia*, **16**, 287–301.

Reed, J. M., Doerr, P. D. and Walters, J. R. (1986) Determining minimum population sizes for birds and mammals. *Wildlife Society Bulletin*, **14**, 255–61.

Reed, J. M., Doerr, P. D. and Walters, J. R. (1988) Minimum viable population size of the red-cockaded woodpecker. *Journal of Wildlife Management*, **52**, 385–91.

Reynolds, J. C. (1985) Details of the geographic replacement of the red squirrel (*Sciurus vulgaris*) by the grey squirrel (*Sciurus carolinensis*) in eastern England. *Journal of Animal Ecology*, **54**, 149–62.

Rice, D. W. and Wolman, A. A. (1971) *The Life History and Ecology of the Gray Whale* (*Estrichtius robustus*), Special Publication No. 3, The American Society of Mammalogists.

Richter-Dyn, N. and Goel, N. S. (1972) On the extinction of a colonizing species. *Theoretical Population Biology*, **3**, 406–33.

Ricker, W. E. (1975) *Computation and Interpretation of Biological Statistics of Fish Populations*, Fisheries Research Board of Canada, Bulletin 191, Ottawa, Canada.

Ripley, B. D. (1987) *Stochastic Simulation*, Wiley, New York.

Robinson, J. G. (1988) Demography and group structure in wedge-capped capuchin monkeys, *Cebus olivaceus*. *Behaviour*, **104**, 202–31.

Roff, D. A. (1974) Spatial heterogeneity and the persistence of populations. *Oecologia*, **15**, 245–58.

Rogers, A. (1966) The multiregional matrix growth operator and the stable interregional age structure. *Demography*, **3**, 537–44.

Rogers, A. (1968) *Matrix Analysis of International Population Growth and Distribution*, University of California Press, Berkeley.

Rogers, A. (1985) *Regional Population Projection Models*, Sage Publications, Beverly Hills.

Roughgarden, J. (1975) A simple model for population dynamics in stochastic environments. *American Naturalist*, **109**, 713–36.

Rowe, W. D. (1977) *An Anatomy of Risk*, Wiley, New York.

Royama, T. (1977) Population persistence and density dependence. *Ecological Monographs*, **47**, 1–35.

Ryan, B. F., Joiner, B. L. and Ryan, T. A. (1985) *Minitab Handbook*, Duxbury Press, Boston.

Sale, P. F. (1977) Maintenance of high diversity in coral reef fish communities. *American Naturalist*, **111**, 337–59.

Sampson, J. F., Hopper, S. D. and James, S. H. (1988) Genetic diversity and the conservation of *Eucalyptus crucis* Maiden. *Australian Journal of Botany*, **36**, 447–60.

Sampson, J. F., Hopper, S. D. and James, S. H. (1989) The mating system and population genetic structure in a bird-pollinated mallee, *Eucalyptus rhodantha*. *Heredity*, **63**, 383–93.

Sarukhan, J. and Gadgil, M. G. (1974) Studies on plant demography: *Ranunculus repens* L., *R. bulbosus* L., and *R. acris* L. III. A mathematical model incorporating multiple modes of reproduction. *Journal of Ecology*, **62**, 921–6.

SAS Institute, Inc. (1985) *SAS User's Guide: Statistics*, Version 5 Edition, SAS Institute, Cary, North Carolina.

Sauer, J. R. and Slade, N. A. (1985) Mass-based demography of a hispid cotton rat (*Sigmodon hispidus*) population. *Journal of Mammology*, **66**, 316–28.

Sauer, J. R. and Slade, N. A. (1986) Size-dependent population dynamics of *Microtus ochrogaster*. *The American Naturalist*, **127**, 902–8.

Sauer, J. R. and Slade, N. A. (1987) Uinta ground squirrel demography: is body mass a better categorical variable than age? *Ecology*, **68**, 642–50.

Saunders, D. A., Rowley, I. and Smith, G. T. (1985) The effects of cleaning for agriculture on the distribution of cockatoos in the southwest of Western Australia, in *Birds of Eucalypt Forests and Woodlands: Ecology, conservation, management* (eds A. Kedst, H. F. Recher, H. Ford and D. Saunders), Surrey Beatty, Chipping North, pp. 309–21.

Saunders, D. A., Hobbs, R. J. and Margules, C. R. (1991) Biological consequences of ecosystem fragmentation: a review. *Conservation Biology*, **5**, 18–32.

Schaal, B. A. and Levin, D. A. (1976) The demographic genetics of *Liatris cylindracea* Michx. (Compositae). *American Naturalist*, **110**, 191–206.

Schaal, B. A., O'Kane, S. L. Jr and Rogstad, S. H. (1991) DNA variation in plant populations. *Trends in Ecology and Evolution*, **6**, 329–33.

Schaffer, W. M. and Kot, M. (1986) Differential systems in ecology and epidemiology, in *Chaos* (ed. A. V. Holden), Princeton University Press, Princeton, pp. 158–78.

Scheuer, E. M. and Stoller, D. S. (1962) On the generation of normal random vectors. *Technometrics*, **4**, 278–81.

Schwaegerle, K. E. and Schaal, B. A. (1979) Genetic variability and the founder effect in the pitcher plant, *Sarracenia purpurea* L. *Evolution*, **33**, 1209–17.

Seber, C. A. F. (1973) *The Estimation of Animal Abundance and Related Parameters*, Griffin, London.

Shaffer, M. L. (1981) Minimum population sizes for species conservation. *Bioscience*, **31**, 131–4.

Shaffer, M. L. (1983) Determining minimum viable population sizes for the grizzly bear. *International Conference on Bear Research and Management*, **5**, 133–9.

Shaffer, M. L. (1985) The metapopulation and species conservation: the special case of the northern spotted owl, in *Ecology and Management of the Spotted Owl in the Pacific Northwest* (eds R. J. Gutierrez and A. B. Carey), U.S. Department of Agriculture, Forest Service, Pacific Northwest Forest and Range Experiment Station, General Technical Report PNW-185, pp. 86–99.

Shaffer, M. (1987) Minimum viable populations: coping with uncertainty, in *Viable Populations for Conservation* (ed. M. E. Soulé), Cambridge University Press, Cambridge, pp. 69–86.

Shaffer, M. L. (1990) Population viability analysis. *Conservation Biology*, **4**, 39–40.

Shaffer, M. L. and Samson, F. B. (1985) Population size and extinction: a note on determining critical population sizes. *American Naturalist*, **125**, 144–52.

Shaw, C. R. and Prasad, R. (1970) Starch gel electrophoresis of enzymes – a compilation of recipes. *Biochemical Genetics*, **4**, 297–320.

Shepherd, J. G. (1982) A family of general production curves for exploited populations. *Mathematical Biosciences*, **59**, 77–93.

Shepherd, J. G. and Cushing, D. H. (1990) Regulation in fish populations: myth or mirage. *Philosophical Transactions of the Royal Society, London B*, **330**, 151–64.

Sherwin, W. D. and Brown, P. R. (1990) Problems in the estimation of the effective size of a population of the eastern barred bandicoot, *Perameles gunnii*, at Hamilton, Victoria, in *Bandicoots* (eds K. Kemper and J. H. Seebeck), Australian Mammal Society, Sydney.

Sherwin, W. B. and Murray, N. D. (1990) Population and conservation genetics of marsupials. *Australian Journal of Zoology*, **37**, 161–80.

Shields, W. M. (1982) *Philopatry, Inbreeding and the Evolution of Sex*, State University of New York Press, Albany, New York.

Simberloff, D. (1986a) Design of nature reserves, in *Wildlife Conservation Evaluation* (ed. M. B. Usher), Chapman and Hall, London, pp. 315–37.

Simberloff, D. (1986b) Are we on the verge of a mass extinction in tropical rain forests?, in *Dynamics of Extinction* (ed. D. K. Elliott), Wiley, New York, pp. 165–80.

Simberloff, D. (1987) The spotted owl fracas: mixing academic, applied, and political ecology. *Ecology*, **68**, 766–72.

Simberloff, D. (1988) The contribution of population and community biology to conservation science. *Annual Review of Ecology and Systematics*, **19**, 473–511.

Simberloff, D. S. and Abele, L. G. (1976) Island biogeography theory and conservation practice. *Science*, **191**, 285–6.

Simpson, G. G. (1952) How many species? *Evolution*, **6**, 342.

Simpson, G. G. (1953) *The Major Features of Evolution*, Columbia University Press, New York.

Skadsheim, A. (1990) A cohort life table for *Gammarus salinus* (Amphipoda). *Oikos*, **57**, 207–14.

Slade, N. A., Sauer, J. R. and Glass, G. E. (1984) Seasonal variation in field-determined growth rates of the hispid cotton rat (*Sigmodon hispidus*). *Journal of Mammology*, **65**, 263–70.

Slatis, H. M. (1960) An analysis of inbreeding in the European bison. *Genetics*, **45**, 275–87.

Slatkin, M. (1974) Competition and regional coexistence. *Ecology*, **55**, 128–34.

Slatkin, M. (1987) Gene flow and the geographic structure of natural populations. *Science*, **236**, 787–92.

Slatkin, M. and Barton, N. H. (1989) A comparison of three indirect methods for estimating average levels of gene flow. *Evolution*, **43**, 1349–68.

Slobodkin, L. B. (1953) An algebra of population growth. *Ecology*, **34**, 513–19.

Smith, A. T. and Peacock, M. M. (1990) Conspecific attraction and the determination of metapopulation colonization rates. *Conservation Biology*, **4**, 320–3.

Sokal, R. R. and Rohlf, F. J. (1981) *Biometry*, (2nd edn), Freeman, San Francisco.

Solbrig, O. T., Newell, S. J. and Kincaid, D. T. (1980) The population biology of the genus *Viola*: I. The demography of *Viola sorona*. *Journal of Ecology*, **68**, 521–46.

Solow, A. R. and Steele, J. H. (1990) On sample size, statistical power, and the detection of density dependence. *Journal of Animal Ecology*, **59**, 1073–6.

Soulé, M. E. (1980) Thresholds of survival: maintaining fitness and evolutionary potential, in *Conservation Biology: An Evolutionary-Ecological Perspective* (eds

M. E. Soulé and B. A. Wilcox), Sinauer, Sunderland, Massachusetts, pp. 151–69.

Soulé, M. E. (1983) What do we really know about extinction?, in *Genetics and Conservation: A Reference for Managing Wild Animal and Plant Populations* (eds C. M. Schonewald-Cox, S. M. Chambers, B. MacBryde and L. Thomas), Benjamin-Cummings, Menlo Park, California., pp. 111–51.

Soulé, M. E. and Simberloff, D. (1986) What do genetics and ecology tell us about the design of nature reserves? *Biological Conservation*, **35**, 19–40.

Soulé, M., Gilpin, M., Conway, W. and Foose, T. (1986) The millenium ark: how long a voyage, how many staterooms, how many passengers? *Zoo Biology*, **5**, 101–13.

Stanley, S. M. (1985) Rates of evolution. *Paleobiology*, **11**, 13–26.

Stanley, S. M. (1987) *Extinction*, Scientific American Library, New York.

Starfield, A. M. and Bleloch, A. L. (1986) *Building Models for Conservation and Wildlife Management*, Macmillan, New York.

Strebel, D. E. (1985) Environmental fluctuations and extinction – single species. *Theoretical Population Biology*, **27**, 1–26.

Strong, D. R. (1986) Density vague population change. *Trends in Ecology and Evolution*, **1**, 39–42.

Sugihara, G., Grenfell, B. and May, R. M. (1990) Distinguishing error from chaos in ecological time. *Philosophical Transactions of the Royal Society, London B*, **330**, 235–51.

Swartzman, G. L. and Kaluzny, S. P. (1987) *Ecological Simulation Primer*, Macmillan, New York.

Sykes, Z. M. (1969) On discrete stable population theory. *Biometrics*, **25**, 285–93.

Szabò, I. (1931) The three types of mortality curve. *Quarterly Review of Biology*, **6**, 462–3.

Taylor, A. D. (1988) Large-scale spatial structure and population dynamics in arthropod predator-prey systems. *Annales Zoologici Fennici*, **25**, 63–74.

Taylor, A. D. (1990) Metapopulations, dispersal, and predator–prey dynamics: an overview. *Ecology*, **71**, 429–33.

Taylor, A. and Hopper, S. D. (1988) *The* Banksia *Atlas*, Australian Government Publishing Service, Canberra.

Templeton, A. R. (1986) Coadaptation and outbreeding depression, in *Conservation Biology: The Science of Scarcity and Diversity* (ed. M. E. Soulé), Sinauer, Sunderland, Massachusetts, pp. 105–16.

Templeton, A. R. and Read, B. (1983) The elimination of inbreeding depression in a captive herd of Speke's gazelle, in *Genetics and Conservation: A Reference for Managing Wild Animal and Plant Populations* (eds C. M. Schonewald-Cox, S. M. Chambers, B. MacBryde and L. Thomas), Benjamin-Cummings, Menlo Park, California, pp. 241–61.

Templeton, A. R. and Read, B. (1984) Factors eliminating inbreeding depression in a captive herd of Speke's gazelle (*Gazella spekei*). *Zoo Biology*, **3**, 177–99.

Templeton, A. R., Shaw, K., Routman, E. and Davis, S. K. (1990) The genetic consequences of habitat fragmentation. *Annals of the Missouri Botanical Garden*, **77**, 13–27.

Terborgh, J. (1974) Preservation of natural diversity: the problem of extinction prone species. *Bioscience*, **24**, 715–22.

Terborgh, J. and Winter, B. (1980) Some causes of extinction, in *Conservation*

Biology. An evolutionary-ecological perspective (eds M. E. Soulé and B. A. Wilcox), Sinauer, Sunderland, Massachusetts, pp. 119–34.

Tigerstedt, P. M. A. (1988) Genetic mechanisms for adaptation: the mating system of Scots Pine, in *Genetics: New Frontiers* (eds V. L. Chopra, B. C. Joshi, R. P. Sharma and H. C. Bansal), Oxford, New Dehli, pp. 317–22.

Tuljapurkar, S. (1989) An uncertain life: demography in random environments. *Theoretical Population Biology*, **35**, 227–94.

Tuljapurkar, S. (1990) Population dynamics in variable environments. *Lecture Notes in Biomathematics*, **85**, Springer-Verlag, New York.

Tuljapurkar, S. D. and Orzack, S. H. (1980) Population dynamics in variable environments. I. Long-run growth rates and extinction. *Theoretical Population Biology*, **18**, 314–42.

Turchin, P. (1990) Rarity of density dependence or population regulation with lags. *Nature*, **344**, 660–3.

Turner, M. E., Stephens, J. C. and Anderson, W. W. (1982) Homozygosity and patch structure in plant populations as a result of the nearest neighbour pollination. *Proceedings of the National Academy of Science*, **79**, 203–7.

Usher, M. B. (1966) A matrix approach to the management of renewable resources, with special reference to selection forests. *Journal of Applied Ecology*, **3**, 355–67.

Usher, M. B. (1969) A matrix model for forest management. *Biometrics*, **25**, 309–15.

Usher, M. B. (1972) Developments in the Leslie matrix model, in *Mathematical Models in Ecology* (ed. J. R. N. Jeffers), Blackwell, Oxford, pp. 29–60.

Usher, M. B. (1986) Wildlife conservation evaluation: attributes, criteria and values, in *Wildlife Conservation Evaluation* (ed. M. B. Usher), Chapman and Hall, London, pp. 1–45.

Usher, M. B. (1987) Effects of forest fragmentation on communities and populations: a review with application to wildlife conservation, in *Nature Conservation: The Role of Remnants of Native Vegetation* (eds D. A. Saunders, G. W. Arnold, A. A. Burbidge and A. J. M. Hopkins), Surrey Beatty, Chipping Norton, New South Wales, pp. 103–21.

Valentine, J. W. (1990) The fossil record: a sampler of life's diversity. *Philosophical Transactions of the Royal Society, London B*, **330**, 261–8.

Vandermeer, J. H. (1975) On the construction of the population projection matrix for a population grouped in unequal stages. *Biometrics*, **31**, 239–42.

Vandermeer, J. H. (1978) Choosing category size in a stage projection matrix. *Oecologia*, **32**, 79–84.

van Groenendael, J., de Kroon, H. and Caswell, H. (1988) Projection matrices in population biology. *Trends in Ecology and Evolution*, **3**, 264–9.

van Hulst, R. (1980) Vegetation dynamics or ecosystem dynamics: dynamic sufficiency in succession theory. *Vegetatio*, **43**, 147.

van Valen, L. M. (1984) Catastrophes, expectations and the evidence. *Paleobiology*, **10**, 121–37.

van Valen, L. M. (1985) A theory of origination and extinction. *Evolutionary Theory*, **7**, 133–42.

Varvio, S., Chakraborty, R. and Nei, M. (1986) Genetic variation in subdivided populations and conservation genetics. *Heredity*, **57**, 189–98.

Verboom, J., Lankester, K. and Metz, J. A. J. (1991) Linking local and regional dynamics in stochastic metapopulation models. *Biological Journal of the Linnean Society*, **42**, 39–55.

Wade, M. J. and McCauley, D. E. (1988) Extinction and recolonisation: their effects on the genetic differentiation of local populations. *Evolution*, **42**, 995–1005.

Walters, C. J. (1985) Bias in the estimation of functional relationships from time series data. *Canadian Journal of Fisheries and Aquatic Sciences*, **42**, 147–9.

Walters and Ludwig (1987) Effects of measurement errors on the assessment of stock recruitment relationships. *Canadian Journal of Fisheries and Aquatic Science*, **38**, 704–10.

Webber, A. W. and Vedder, A. (1983) Population dynamics of the Virunga Gorillas: 1959–1978. *Biological Conservation*, **26**, 341–66.

Wegner, J. F. and Merriam, G. (1979) Movements by birds and small mammals between a wood and ajoining farmland habitats. *Journal of Applied Ecology*, **16**, 349–57.

Werner, P. A. and Caswell, H. (1977) Population growth rates and age versus stage-distribution models for teasel (*Dipsacus sylvestris* Hud.). *Ecology*, **58**, 1103–11.

Whitehead, P. J. and Tschirner, K. (1990) Magpie goose, *Aseranas semipalmata*, nesting on the Mary River floodplain, Northern Territory, Australia: extent and frequency of flooding losses. *Australian Wildlife Research*, **17**, 147–57.

Whittaker, R. H. and Goodman, D. (1979) Classifying species according to their demographic strategy. I. Population fluctuations and environmental heterogeneity. *American Naturalist*, **113**, 185–200.

Wiens, D., Nickrent, D. L., Davern, C. I., Calvin, C. L. and Vivrette, N. J. (1989) Developmental failure and loss of reproductive capacity in the rare palaeoendemic shrub *Dedeckera eurekensis*. *Nature*, **338**, 65–7.

Williamson, M. H. (1959) Some extensions of the use of matrices in population theory. *Bulletin of Mathematical Biophysics*, **21**, 13–17.

Wilson, E. O. (1988) The current state of biological diversity, in *Biodiversity* (eds E. O. Wilson and F. M. Peter), National Academy Press, Washington, pp. 3–18.

Wilson, E. O. and Willis. E. O. (1975) Applied biogeography, in *Ecology and Evolution of Communities* (eds M. L. Cody and J. M. Diamond), Harvard University Press, Cambridge, Massachusetts, pp. 522–36.

Wold, H. (1955) *Random Normal Deviates*, Cambridge University Press, Cambridge.

Wolfenbarger, D. O. (1946) Dispersion of small organisms. *American Midland Naturalist*, **35**, 1–152.

Woodruff, D. S. (1989) The problems of conserving genes and species, in *Conservation for the Twenty-first Century* (eds D. Western and M. C. Pearl), Oxford University Press, New York, pp. 76–88.

Workman, P. L. and Niswander, J. D. (1970) Population studies of southwestern Indian tribes. II. Local genetic differentiation in the Papago. *American Journal of Human Genetics*, **22**, 24–49.

Wright, S. (1931) Evolution in Mendelian populations. *Genetics*, **16**, 167–76. In *Evolution, Selected Papers* (ed. W. B. Provine), 1986, University of Chicago Press, Chicago.

Wright, S. (1939) Statistical genetics in relation to evolution. *Exposes de Biometrie et de Statistique Biologique*, Herman, Paris. In *Evolution, Selected Papers* (ed. W. B. Provine), 1986, University of Chicago Press, Chicago.

Wright, S. (1948) On the roles of directed and random changes in gene frequency in the genetics of populations. *Evolution*, **2**, 279–94.

Wright, S. (1951) The genetical structure of populations. *Annals of Eugenics*, **15**, 323–54.

Wright, S. J. (1977) Inbreeding in animals: differentiation and depression, in *Evolution and the Genetics of Populations, 3. Experimental Results and Evolutionary Deductions* (ed. S. Wright), University of Chicago Press, Chicago, pp. 44–96.

Young, T. P. (1984) Comparative demography of semelparous *Lobelia telekii*, and iteroparous *Lobelia kenitensis* on Mount Kenya. *Journal of Ecology*, **72**, 637–90.

Zeckhauser, R. J. and Viscusi, W. K. (1990) Risk within reason. *Science*, **248**, 559–64.

Zimmerman, B. L. and Bierregaard, R. O. (1986) Relevance of the equilibrium theory of island biogeography and species-area relations to conservation with a case from Amazonia. *Journal of Biogeography*, **13**, 133–43.

Index

All references in *italics* represent figures, those in **bold** represent tables.